T0188863

Re-Visioning Europe

Re-Visioning Europe

Frontiers, Place Identities and Journeys in Debatable Lands

Ullrich Kockel
University of Ulster, UK

© Ullrich Kockel 2010
Softcover reprint of the hardcover 1st edition 2010 978-1-4039-4122-0

First published 2010 by
PALGRAVE MACMILLAN

Palgrave Macmillan in the UK is an imprint of Macmillan Publishers Limited, registered in England, company number 785998, of Houndmills, Basingstoke, Hampshire RG21 6XS.

Palgrave Macmillan in the US is a division of St Martin's Press LLC, 175 Fifth Avenue, New York, NY 10010.

Palgrave Macmillan is the global academic imprint of the above companies and has companies and representatives throughout the world.

Palgrave® and Macmillan® are registered trademarks in the United States, the United Kingdom, Europe and other countries.

ISBN 978-1-349-52060-2   ISBN 978-0-230-28298-8 (eBook)
DOI 10.1057/9780230282988

This book is printed on paper suitable for recycling and made from fully managed and sustained forest sources. Logging, pulping and manufacturing processes are expected to conform to the environmental regulations of the country of origin.

A catalogue record for this book is available from the British Library.

A catalog record for this book is available from the Library of Congress.

10   9   8   7   6   5   4   3   2   1
19   18   17   16   15   14   13   12   11   10

Transferred to Digital Printing in 2014

Da liegt Europa. Wie sieht es aus?
Wie ein bunt angestrichnes Irrenhaus.
Kurt Tucholsky: "Europa" (1932)

Dat gah uns woll op ole Daag!
Martje Flor

Contents

List of Figures and Tables

Figures

Tables

Preface

Questions about where and what 'Europe' is have featured prominently in the social sciences and humanities as well as in political discourse since 1989, not least in the US government's recent distinction between an 'old' and a 'new' Europe, the eastward expansion of the EU and the drafting of a European Constitution. Using journeys in pursuit of a vision as a metaphor, this book explores key issues for contemporary Europe and its future development from an interdisciplinary perspective grounded mainly in European ethnology, cultural anthropology, human geography and political economy.

These journeys begin with an exploration of Ulster, one of the 'debatable lands' of historical as well as contemporary Europe. Originally, the term 'debatable lands' was used in the sixteenth century for areas of disputed sovereignty along the Anglo-Scottish border. Used more widely since the Romantic period, it has since come to designate not only contested geographies across the world but also disputes in the sphere of intellectual, political or artistic development, especially with regard to place identities. In the absence of clearly defined boundaries, 'debatable lands' offer spaces in which the unexpected may unfold. In five substantive chapters, the book offers a synthesis dealing with migration within and into Europe; frontiers and boundaries; heritage and tradition; socio-economic structures, processes and change; and, finally, the role of ethnology in education and cultural practice. The book is framed by personal reflections on changing visions of Europe, suggesting a fresh envisioning of a dis-placed continent.

The basic framework for this book is provided by my 2001 inaugural lecture at the University of the West of England, Bristol, subsequently published as 'EuroVisions: Journeys to the Heart of a Lost Continent', *Journal of Contemporary European Studies* 11(1), 2003, 53–66 (http://www. informaworld.com), extracts of which are reproduced here with permission by Taylor & Francis.

Material from the following texts, revised and updated as appropriate, has been used with permission of the publishers, which is herewith gratefully acknowledged:

'Nationality, Identity, Citizenship: Reflections on Europe at Drumcree Parish Church', *Ethnologia Europaea* 29(2), 1999, 97–108.

'Protestantische Felder in katholischer Wildnis: Zur Politisierung der Kulturlandschaft in Ulster', in R. Brednich, A. Schneider and U. Werner eds, *Natur – Kultur: Volkskundliche Perspektiven auf Mensch und Umwelt* (Münster: Waxmann 2001), 125–38.

'Heimat als Widerständigkeit: Beobachtungen in einem Europa freischwebender Regionen', in S. Götsch and C. Köhle-Hezinger, eds, *Komplexe Welt: Kulturelle Ordnungssysteme als Orientierung* (Münster: Waxmann 2003), 167–76.

'Von der Schwierigkeit, "Britisch" zu sein: Monokulturelle Politik auf dem Weg zur polykulturellen Gesellschaft', in C. Köck, A. Moosmüller and K. Roth, eds, *Zuwanderung und Integration: Kulturwissenschaftliche Zugänge und soziale Praxis* (Münster: Waxmann 2004), 65–81.

'"Authentisch ist, was funktioniert!" Tradition und Identität in drei irischen Städten', in S. Göttsch, W. Kaschuba and K. Vanja eds, *Ort – Arbeit – Körper. Ethnografie europäischer Modernen* (Münster: Waxmann 2005), 127–34.

'Heritage versus Tradition: Cultural Resources for a New Europe?' in M. Demossier, ed., *The European Puzzle: The Political Structuring of Cultural Identities at a Time of Transition* (Oxford and New York: Berghahn 2007), 85–101.

'K(l)eine Deutschlande: Heimat und Fremde deutscher Einwanderer auf den Britischen Inseln', in E. Tschernokoshewa and V. Granssow eds, *Beziehungsgeschichten. Minderheiten – Mehrheiten in europäischer Perspektive* (Münster: Waxmann 2007), 188–202.

'Editorial', *Anthropological Journal of European Cultures* 17(1), 2008, 1–4.

'Putting the Folk in Their Place: Tradition, Ecology, and the Public Role of Ethnology', *Anthropological Journal of European Cultures* 17(1), 2008, 5–23.

'Liberating the Ethnological Imagination', *Ethnologia Europaea* 38(1), 2008, 8–12.

This book draws on fieldwork in Britain, Ireland, Germany, Denmark, Sweden, Finland, Estonia, Lithuania, Poland, Romania, Slovenia, Italy, France and Spain, carried out between 1989 and 2009. It would be impossible to name everyone 'out there' who has contributed to this research in one way or another – agreeing to be interviewed, offering introductions, guiding my initial explorations of new places, discussing my impressions and interpretations and so much more. Thank you all – I hope that you can recognise yourselves in the story I have tried to tell here.

As all academics, I owe a huge debt to my mentors, colleagues and students. Where individuals have made a particular contribution to shaping the ideas in this book, this is acknowledged in the Notes. I would also like to thank the editorial staff at Palgrave Macmillan for commissioning this work and accompanying its circuitous progress with admirable patience, and the anonymous reviewers for suggesting useful improvements (even if I did not implement them all).

This book is dedicated to Máiréad Nic Craith, whose idea it was and who not only read and improved several drafts but has also sustained me spiritually throughout all my winding journeys. What I made of the visions I encountered en route is, of course, entirely my own responsibility.

1
Setting Out: *Europe – A Winter's Tale ...*

Once upon a time, not all that long ago, there was a place called 'Europe', which some of us may still remember. Like other parts of the world, it had its share of problems, but most people were, nonetheless, happy enough there, while many from elsewhere have long been eager to move to Europe, even as we are told that there may not be, nor ever have been, such a place at all. Until not so long ago, it seemed to be a fairly obvious matter where and what this Europe was. The Danes saw it as the area between their province of Sønderjylland and the Dolomite mountains in northern Italy;[1] to the English it was the intriguingly barbarian frontier beyond the homeland of their Norman colonisers; for the Russians it was a kind of cultural *alter ego*. The French and Germans would expect Europe to be a place where, at long last, they might live together in peace. And some incurable romantics regarded Kakania, the Austro-Hungarian monarchy, as the prototype for a multicultural 'Europe'. While it may have meant rather different things to different people, there was at least a consensus that it did exist somewhere. But that consensus evaporated in the final decades of the twentieth century.

As climate change is beginning to turn the seasons, at least in the part of the world where I live and write, into a kind of perennial lukewarm winter, I wonder whether there would not be more important matters to consider than the elusive identity of a place called 'Europe' – after all, a place that, according to fashionable opinion both public and academic, does not even have any existence, let alone justification, outside the imagination of vested interests whose political agenda is rather dubious. However, having never been a 'dedicated follower of fashion' (The Kinks 1966), I shall stubbornly persist in my quest to find this Europe, a quest that started more than ten years ago with reflections on issues raised by the Amsterdam Treaty.

It might be tempting for the decimal mind to take that 1999 treaty as a time horizon for this book, written ten years later. Alternative time frames could be provided by other key dates in the history of European integration – 1992, 1986, 1981, 1979, 1973 and so on, all the way back to the Treaty of Rome, 1957, and indeed beyond. Initially, I had settled for 1979, which marked the conjunction of several key events, including the first general elections to the European Parliament, the creation of a European Monetary System and the second 'oil crisis'; however, an anonymous reviewer commenting on my application for research leave to complete this book thought that my choice of date was quite 'arbitrary', implying the year 1979 had no particular relevance for Europe, other than perhaps in my own mind. In deference to the time-honoured tradition of peer review, I shall therefore refrain from the use of any arbitrary time frame until someone can suggest a non-arbitrary one. The reader may find the lack of a consistent historical frame of reference puzzling and disorienting, but I am actually rather grateful to that reviewer for liberating my inquiry from the spurious constraints of calendar time. Indirectly, this act of liberation has also reinforced my intuition that 'Europe' is not limited to the European Union (EU), however much the two, nowadays, tend to be conflated in colloquial speech (and the minds of peer reviewers). My concern in this book is therefore not with any 'everyday Europe' – as in how does the EU impact on, or how is it negotiated in, everyday life – but with the larger, deeper Europe, the Europe that existed long before any EU and will, climate change and other factors permitting, be here long after that EU has followed the Holy Roman Empire, the Hanseatic League and other precursors into oblivion. If this Europe is not bigger and deeper than the EU, it is not at all – even if every country that meets the criteria (and some that do not) had joined it long ago. If such a Europe does indeed exist in any meaningful and sustainable way, then its analysis should not be forced into a straightjacket of dates and historical timelines.

Nevertheless every journey needs to start from some place, and at some time. My exploration of Europe starts in Hamburg in the late 1970s, not because that was a significant time for Europe (which it was all the same) but because this was when and where I first defined myself as a European. Arbitrary as this may be, to do otherwise would be to deny that the times and places that have shaped us in our youth have any bearing on the interests and ideas we develop later in life.

Long before the new cosmopolitanism became the flavour of the moment in academia, it had been fashionable in the coffee houses of Western Central Europe (at least). Thoroughly inculcated with the core

values of 'the Western World' – *verwestlicht*, as some might say – many
in my generation, as they were coming of age, identified themselves
deductively as human beings first, then cosmopolitans, then Europeans;
only then, often an afterthought, came identification with the nation
state. This was done in deliberate contradistinction to a generation of
parents and grandparents perceived as having taken parochial myopia
to its devastating conclusion – even where that may have happened by
default rather than as a result of individual intent. At the same time, it
was a reaction against the spurious internationalism of the Communist
bloc that was recognised as masking oppressive, totalitarian regimes.
Few in the West were aware that among our cohort on the other side of
the Iron Curtain, the resistance to these regimes very often took the
form of inductive identifications via the revival of traditions (see, e.g.
Čiubrinskas 2000).

55 million votes for Europe

The 1960s had been a strange time. To a boy growing up in a country
that ended ominously at dotted lines in areas one could not easily
enter,[2] the decade had been full of television images in black and white:
John and Bobby Kennedy, Martin Luther King, the massacre at My Lai,
Jan Palach's burning body and the Soviet tanks in Prague, the Burntollet
Bridge ambush of a civil rights march, the killing of Benno Ohnesorg in
Berlin at a rally against the Shah of Persia – a shot that would reverberate
throughout Germany for years to come, the wars in the Middle East
and, again and again, Vietnam.

By the 1970s, television had acquired colour but the images were
no better: anti-war protesters shot dead at Kent State University, Ohio;
Israeli athletes murdered during the Olympic Games in Munich; the
NATO allies Turkey and Greece almost at war with each other over Cyprus;
a general strike toppling the power-sharing government in Northern
Ireland; international power politics showing its face at our school as
children of Chilean refugees joined our classes; carnations placed in the
barrels of Portuguese guns add a lighter touch to the vision of a period
that was about to end with the death of Rudi Dutschke from a bullet
fired by an assassin back in 1968.

No, the so-called post-war world was anything but peaceful. And into
that atmosphere, chilled to the bone by the 'leaden time',[3] was projected
the vision of a stable European peace, *eine europäische Friedensordnung*,
an ideal that had no difficulty attracting the support of impressionable
first-time voters in the first European elections of 1979. The two world

wars had destroyed German politician Gustav Streseman's dream of a liberal Europe of peaceful coexistence; the Free Democratic Party (FDP) took up that baton for the forthcoming elections (Figure 1.1). They said only those who acted responsibly ought to be given responsibility. And 55 million 'votes for Europe', cast before a European Community was even heard of, surely had to be a persuasive tally in their favour.

It did not occur to us to ask how many of those 55 million dead might actually have voted for the Europe of this or any other vision. The appeal was to rational argument, and yet the campaign rode on a wave of anti-war emotion that had been swelling in the country ever since the Easter marchers of the late 1950s (cf. Otto 1977). It was to peak with the public debate over NATO's Dual-Track strategy that led to the 'Euro-missile crisis' of the mid-1980s, and did not ebb away until the 1990s after the fall of the Berlin Wall. The campaign worked, the FDP won over 1.6 million votes – more than in any subsequent European election until 2004 – and returned four delegates to the European Parliament (MEPs).[4]

The 'Europe-Blanket'

As the citation of Gustav Streseman in the FDP advert shows, the idea of connecting a vision of Europe to a vision of sustainable peace and goodwill is not new. Nor is it the intellectual property of this or any other political party. Every journey has to start from somewhere and one of the journeys to Europe started with the project to appease two quarrelsome neighbours, France and Germany. Historically, one of the causes for their frequent disputes appears to lie underground: the coal and steel resources of Alsace-Lorraine and the Saarland. To settle these disputes, the Treaty of Paris in 1951 created a European Coal and Steel Community. The Treaty of Rome in 1957 established its better-known partner, the European Economic Community (EEC).

To win the hearts and minds of the people who, until not so long ago, had been whipped into hating each other, popular culture was marshalled. However, in the spirit of post-war frugality, this did not mean the usual dust gatherers distributed for commemorative purposes. Instead, the new reality of 'Europe' was publicised on more practical items of everyday use. One of these items I have grown rather fond of over the years, as it has accompanied my perambulations along the length of the Iron Curtain, from Göttingen to Schleswig-Holstein to Franconia, back north, and ultimately further afield: the 'Europe-blanket' with its multilingual label (Figure 1.2).

55 MILLIONEN STIMMEN FÜR EUROPA.

Gustav Stresemann entwickelte 1929 vor dem Völkerbund seine Traumvorstellung von Europa. Einem Europa des Miteinander-Sprechens, des Ausgleichs, der Verständigung. Ein liberales Europa.

Dieser Traum wurde zerstört. Die Zeit war noch nicht reif für seine Ideen.

Jetzt, 50 Jahre später, hat Europa seine zweite Chance. Europa soll eins werden. Denn nach zwei Weltkriegen haben die Menschen in Europa gelernt, wohin Nationalstaaterei, Unverständnis und Ignoranz gegenüber den europäischen Nachbarn führt.

Daß man Gegensätze überbrücken kann, zeigt das Beispiel der europäischen Liberalen: Sie stellen sich als eine Partei zur Wahl.

Früher als die übrigen Parteigruppierungen haben Vertreter der liberalen Parteien aus England, Frankreich, Dänemark, Italien, den Benelux-Ländern und Deutschland ihr Programm für Europa entwickelt. Es wurde beschlossen, als die Konservativen und Sozialisten noch um Führungsrollen und Wahlkampfsymbole stritten.

Die Liberalen sind für ein friedliches Europa. Einen Krieg zwischen den Ländern der europäischen Gemeinschaft kann man sich heute schon nicht mehr vorstellen.

Die erfolgreiche Aussöhnungspolitik der Liberalen macht auch die Gefahr eines Krieges mit anderen Ländern geringer. Deshalb muß das ganze Europa die Entspannungspolitik nach Osten fortsetzen und Partner im Nord - Süd - Dialog werden.

Denken Sie daran, wenn Sie am 10. Juni Europa wählen. Denn Verantwortung tragen kann nur, wer verantwortlich handelt.

ÜBERLASSEN SIE EUROPA NICHT DEN ANDEREN.

DIE F.D.P. IST MITGLIED DER FÖDERATION DER EUROPÄISCHEN LIBERALEN DEMOKRATEN: DER ELD

An die Föderation der ELD, 1, Boulevard de l'Empereur, B-1000 Bruxelles. Helfen Sie durch Ihre Veröffentlichungen weiter, die Notwendigkeit der europäischen Einheit zu fördern.

NAME

STRASSE

PLZ/ORT

F.D.P.
Europa liberal

★ ★ ★
★ ★ ★
★ ★ ★

Figure 1.1 A German liberal party campaign advert for the first elections to the European Parliament, 1979
Source: Reproduced courtesy of Friedrich-Naumann-Stiftung, Archiv des Liberalismus.

Figure 1.2   Label of the 'Europe-Blanket', 1957
Source: Personal archive.

With time, as television became more sophisticated, additions to the arsenal included popular TV shows such as *Jeux Sans Frontières*, a fun-and-games competition involving teams from different European countries that had started its life in France as *Intervilles*, running internationally from 1965, in Germany under the title *Spiel ohne Grenzen*, as a Eurovision broadcast organised by the Italian company RAI. A year earlier, the federation of German regional broadcasting companies had launched its 'great international quiz', moderated by the highly popular actor Hans-Joachim Kulenkampff. This Saturday prime-time show would run for well over 20 years, until 1987, making it one of the most successful, longest-running programmes on German television. Its title was *Einer wird gewinnen*,[5] which can be abbreviated as 'EWG', and this, in German, is the abbreviation of the EEC.

At the end of the 1970s, folk groups from abroad – mostly Ireland and Scotland – liked touring West Germany, but when we sat together for an *après-tour* session at the *Zwiebel* in Hamburg, that legendary Irish pub that preceded by several decades the fad for Irish-themed pubs in the 1980s, their gaze more often than not was directed east, to Berlin – and beyond. At concerts as well as in the pubs, German folk music fans (and others) liked to sing songs like the 1830s Irish nationalist anthem, *A Nation Once Again*. Expressing one's national sentiments was no easy feat for a generation deeply immersed in the anti-fascist, anti-war and Green movements, but Ireland, more than any other Celtic country perceived as Third World colony on our doorstep, was recognised as a legitimate arena and context for this. That may be one reason why Irish music was the most popular of all the various folk traditions that helped stimulate the revival of regional languages and promote the interweaving of folk music and protest movements. That process penetrated the Iron Curtain in the 1970s and led to the rise of popular bands like the *Folkländer* in the German Democratic Republic, who drew inspiration from Irish and Scottish music to develop their own, cautiously countercultural repertoire. Here we glimpse a Europe of sorts (and one we may come back to later) that stretches much further than the European blanket with its four languages.

The loss of Europe

Europe – but where is it now? Even St Benedict, declared a patron saint of Europe by Pope Paul VI in 1964, was unable to defend his charge against the assault by constructivists and deconstructionists who, before long, set about disposing of Europe discursively. The deconstructionists revealed our familiar ideas and philosophical approaches as shamefully Eurocentric. I am not sure whether anyone dared to ask how they possibly could be 'Eurocentric' at all if there was no such thing as 'Europe' in the first place, but this might have been an unfashionable comment to make in the circumstances. Eurobashing had become a well-established intellectual ritual. Enter the constructivists, who see it all as invented, typically by powerful political interests or by everyone else's 'false consciousness', and Europe was quickly twisted to become a non-place. The critics made so much noise about the novelty of their approach that they failed to notice how they themselves were only reinventing a philosophical wheel that had revolved in ancient Athens, India and China, and probably elsewhere, and that their innovative analysis was deeply rooted in the maligned 'European tradition'. Of course, there are various visions[6] of Europe, and each of these rests on its own set of premises. As the medieval historian Michael Borgolte (2005: 124) points out, any historical concept or representation of Europe is a construct that will only hold if you accept its premises. Whether you buy into this argument depends less on whether or not there is a Europe, or what it looks like, than on whether you are an idealist or a realist at heart – not just philosophically speaking.

Arguably, Europe can be easily overlooked on a rapidly spinning globe. As the Irish literary critic Declan Kiberd (2005: 255) has observed, 'to the Japanese Europe now appears as a tiny, open-air boutique at the fag-end of Asia'. In this light, the growing interest in Eurasia in Anglophone social anthropology (e.g. Hann 2006) is perhaps not surprising, although why Eurasia should be regarded as less of a construct than Europe remains a puzzle. Already some two-and-a-half millennia ago, the Greek historiographer Herodotus criticised the conceptual division of the known world of his day into three continents, because he saw Europe, Asia and Libya (i.e. Africa) as a single land mass.

In certain ways, due to its colonial expansion in the past, Europe is everywhere. Its colonial past is one reason for the negative image Europe has acquired in academic and popular perception in recent decades, an image frequently invoked in proposals to accelerate its conceptual demise. And yet, it is sometimes said that 'one person's coloniser may be another person's entrepreneur!' Peripheral regions, often referred to as internal colonies, know a thing or two about this dilemma. In some ways, contemporary Europeans may be regarded as members of postcolonial communities, or as 'projections of global diversity within the European sphere' (Balibar 2004: 8). Arguably, 'the remoteness which Europe once shaped is now reshaping Europe' (Kiberd 2005: 255), and a non-Eurocentric Europe may be emerging as a result of processes of creolisation while cosmopolitanism may create new communities. Its history of colonisation and migration has meant that Europe, be it invented or not, has become increasingly difficult to delineate. In this situation, some old conceptual boundaries may be revived. Elias Canetti wrote in his autobiography that people in his native village on the lower Danube referred to someone who went up the river to Vienna as someone who was 'going to Europe. Europe began where the Turkish Empire had once ended' (Canetti 1999: 6). A century later, Europe is once again – or still – defined in contradistinction to the Islamic world. At a time when the EU has just acquired its first indigenous Muslim minority, in its new member state of Bulgaria, and Turkey itself is applying to join the EU, this makes less sense than ever. Creating Europe has always been, as the Lithuanian poet Thomas Venclova has pointed out, a task full of uncertainty and risk. He was writing about Vilnius, a place permanently on the periphery and in the frontier – eccentric, capricious, irregular – a city with a strange past and which breaks the rules of logic and probability (Venclova 2006: 242). It seemed an obvious place to visit at some stage on my journeys in search of Europe. By the time I made it there in 2007, Laimonas Briedis was already writing his excellent book on that 'city of strangers' (Briedis 2008).[7]

The American anthropologist Carol Rogers (1997: 719) observed some time ago that studies of Europe are 'in a state of considerable disarray. No one is quite sure even where the boundaries of Europe now lie'. Beyond 'European Studies' as an undergraduate subject, which at most universities is keenly focused on EU integration, the confusion persists. Some Europeanists now see culture and 'related forms of irrationality' as providing some kind of 'black-box explanation' for everyday patterns and practices that seem to make little or no sense otherwise. Just as archaeologists allegedly label any material item they cannot explain as 'religious object', so Europeanists across the disciplines declare as 'cultural' anything they can neither understand nor approve of otherwise (especially if it happens in Central and Eastern Europe).

But what is the point of looking for Europe? Did not Europe, if it ever existed, die long ago? One hears this diagnosis frequently, usually uttered in the same breath as the break-up of the Austro-Hungarian Empire or, more often, the Holocaust. In either case, the reference to Europe is in fact to a 'Central European' culture, perceived as characterising what in an Irish translation of a much-cited essay by Milan Kundera (1990) is called *Croí na hEorpa*, 'heart of Europe'. While a certain nostalgia for the Golden Age of the Habsburg monarchy may appear in some texts addressing the question of Central Europe, most authors have a more differentiated view of the region and its history than critics from outside give them credit for.[8] The identity of the region tends to be closely associated with its *shtetl* culture. The brutal extinction of this culture in the Holocaust is recognised – not just by academics (e.g. Schlögel 2002: 43) and popular writers hailing from elsewhere but, more importantly, also in the poetry and prose of authors from the region, such as Johannes Bobrowski or Czesław Miłosz – as a key factor in the destruction of (Central) Europe as a specific historical context. In post-1968 Paris, Jewish intellectuals – '[i]nspired by symbolic, not historical truth' – created a kind of 'Vilna on the Seine', reminiscent of 'the old Jewish communities that once stood along the rivers Neris and Neman and across Lithuania's thick forests' (Friedlander 1990: 5–6). Some of them, for example Lévinas, have had major influence on twentieth-century thought, and thus have arguably contributed to the survival and revival of an older Europe in spite of the gas chambers. The horror of the Holocaust is terrible proof that, whatever constructivists might say,

Europe is emphatically not just an invention. To pronounce Europe dead because of it amounts to a refusal to engage with the dark side of that Europe.

The homepage of the Berlin-based company *European Exchange* quotes Susan Sontag's suggestion that Europe may be far from dead: 'It may be truer to say that Europe has yet to be born.'[9] In a later essay on 'The Idea of Europe (One More Elegy)', Sontag (2003: 289) argues, however, that the territory Europe now occupies is ever shrinking, and that 'increasing numbers of its citizens and adherents will understand themselves as émigrés, exiles, and foreigners' – strangers in our own land. There is more than a grain of truth in this analysis. Europe does exist even if late-, post- or hyper-modernity, which ever it is we may be gradually emerging from, has successfully dismantled our cosy certainties. Shrouded in a mist of discourse, the maps burnt on the constructivist *falò delle vanità*,[10] this Europe is now almost unknowable. It has become a non-place. The ancient Greeks might have called it οὐτόπος – utopia.

So how should we approach this non-place? More to the point, how will we know whether what we find and see has anything to do with Europe at all? Such is the epistemological quagmire that we have allowed ourselves to be led into by all that discourse. The only analysis permitted is one that dissects how a particular construct has been constructed, by whom and for what. Everything else is essentialism. Do not dare ask the constructivists how and why they construct those metaphysical forces that allegedly construct everything; the infinite regress of their approach might become only too evident.

Se una notte d'inverno un viaggiatore ...[11]

Living now on an archipelago off the north-west European mainland, I often hear people talking about 'going to Europe' when they are merely crossing the English Channel – much as the Bulgarian villagers of Canetti's childhood would travel up the Danube 'to Europe'. Some 20 years ago, I began to study this 'Europe' that I had subscribed to enthusiastically some ten years earlier, when I had joined the European Movement for a time. That early research still took Europe, as a geographical and historical entity, largely for granted. However, by the time it had been worked into my first book on European integration (Kockel 1999a), the assumptions underpinning what – with the benefit of hindsight – seems a rather innocent world view had crumbled. At the same time, other, more personal parameters had consolidated.

My self-imposed island exile looked ever more likely to turn into a permanent arrangement and this raised questions of perspective, not just because I had long aligned myself conceptually with European ethnology and cultural anthropology.

In my 2001 inaugural lecture at the University of the West of England, Bristol, I tried to set some pointers for an exploration of the wider Europe, drawing on my earlier research and some debates current at the time. That lecture provided the initial impetus for this book. The lecture had been divided into visions corresponding to journeys (both actual and metaphorical). One of these – the economy – became a major concern in my second book on European integration (Kockel 2002). But there were changes that shaped the direction the present work was taking. Moving to what George Bernhard Shaw in 1904 called 'John Bull's Other Island' was one of these changes, which again entailed a shift in perspective. For some time, this project took on a very different, in a sense 'genre-busting' format and became more about an actual rather than a purely metaphorical, discursive quest for a lost Europe and its Europeans. In retrospect, I can trace this shift of emphasis to symposia that I organised: the Research Seminars in European Ethnology, funded by the Economic and Social Research Council (ESRC) in 2001–3, and a 2005 mini-conference on cultural encounters at the eastern borders of the EU, funded by the University Association for Contemporary European Studies (UACES). However, as a substantive new direction for my research, this perspective has only begun to take shape since I joined the University of Ulster. I am grateful to the Faculty of Arts there for granting me leave during the winter semester 2009/10, which finally allowed me to finish this book, travelling the landscapes of the mind on many a winter's night, retracing my journeys in search of Europe. To fit the project within the given time frame for completion, some of the more experimental aspects had to be postponed for another occasion, so that the finished work now, once again, looks more like the original proposal – although not quite ...

While this book is grounded in extensive and intensive ethnographic research carried out in various locations across Europe, it does not offer any detailed ethnography of the EU or of the effects of EU policies, nor does it concentrate on detailed local studies of 'Europe from below', although the early chapters in particular provide some ethnographic sketches. My approach has been to use critical reflection on ethnographic encounters to see if the contours of a wider Europe that could inform the design of a future research programme might be discerned. The chapter sequence follows the five visions and journeys that I outlined in 2001, with Chapters 2 and 3 drawing extensively on ethnographic work largely completed before I came to Ulster, and continuing to explore themes and issues raised in my two previous books on European integration. Chapters 4 and 5 move on in different but related directions – deeper into Europe as a place of cultural realisation, and towards a radical questioning of our received ways of knowing and making the world, including Europe. Chapter 6 picks up some of those cues and tries to set pointers for a research programme; it thus connects with the fifth vision journey of 2001, which looked at higher education at the crossroads and the role of ethnology in that context. That topic requires more extensive and intensive treatment than is possible in the present book.

Ethnographic fieldwork has been a defining characteristic of European ethnology for a long time. The narrative in this book is based on extensive ethnographic research, but it does not provide highly detailed ethnography of an everyday Europe. Such ethnographies do exist, and had I continued on the original track, my book might well have joined their ranks. Instead, I am trying to capture here glimpses of a Europe I have encountered over the past few years, 'confronting something and not quite knowing what it is' (Calvino 1992: 9). An early advocate of fieldwork was Wilhelm Heinrich Riehl, a professor at Munich University in the mid-nineteenth century, with his approach of *Erwanderung*, of getting to know the field by rambling and roaming (Zinnecker 1996; Girtler 2004).

On my ethnographic meanderings around Europe, I have often felt spiritual kinship with a poet whose work I first picked up in that most inconsequential of years, 1979. Heinrich Heine, a contemporary of Riehl, composed the epic poem 'Germany: A Winter's Tale'; in it he described a journey all over Germany in search of a vision of that country, which he sets out in the first sequence, 'Caput I'. There we also encounter Europe:

> *Die Jungfer Europa ist verlobt*
> *Mit dem schönen Geniusse*
> *Der Freiheit, sie liegen einander im Arm,*
> *Sie schwelgen im ersten Kusse.*
>
> *Und fehlt der Pfaffensegen dabei,*
> *Die Ehe wird gültig nicht minder –*
> *Es lebe Bräutigam und Braut,*
> *Und ihre zukünftigen Kinder!*[12]

Embracing the Spirit of Freedom to whom she is betrothed, Europe indulges in a first kiss; the wandering poet salutes bride and groom, and their future children, even if the authorities – represented here by the clergy – may not give them their blessing. That vision of Europe seems worth a few journeys. Bear with me as – proceeding from the corner of Europe where I live right now and gradually moving further afield – I ramble and roam some rather nebulous fields and debatable lands, pausing here and there to listen for the heart of Europe beating through the mists of discourse.

2
First Journey – In the Frontier: *Balancing on Borderlines*

In his introduction to Ivan Olbracht's novel *The Sorrowful Eyes of Hannah Karajich*, Miroslav Holub tells a story about a Hasidic Jew who was asked how many countries he had seen (Olbracht 1999: vii):

> Well, he says, I was born in Austria-Hungary, I was married in Czechoslovakia, I was widowed in Hungary, and now I'm trying to make ends meet in the Soviet Union. Been quite a traveller then, haven't you? Not at all. I never moved a step from Mukachevo.

Mukachevo, where Olbracht's novel is set, lies in Sub-Carpathian Ruthenia, which nowadays is part of Ukraine. The story highlights that you do not need to migrate to find yourself on the other side of a national border, whether you like it or not.

The end of the old European order has been brought about by a combination of large-scale population movements and a resurgence of ethnic nationalisms, and European ethnologists have been pondering the idea of a 'virtual ethnicity' (Köstlin 1996: 178): 'If history fills our memory, ... a virtual identity based on expert stories will be thinkable.' But is this not what happened throughout the history of nationalism? Is not national identity the virtual identity par excellence?

The late Frank Wright (1987) applied the notion of 'frontier' in his analysis of Ulster, juxtaposing it to the 'metropolis' in what is essentially a modified core–periphery perspective. His view of the 'frontier' was similar to the vision of the American historian Frederick Turner, who characterised it as 'a region of opportunity' (Hannerz 1997: 538). In late twentieth-century terms, the 'frontier' is an open field where identities can be de- and re-constructed. Ethnic frontiers are characterised by two or more 'separate and competing claims which are made simultaneously on the loyalty of all the people on the territory' (Morrow 1994a: 341). Where rival groups have achieved a 'critical mass', parallel ethnic institutions have developed. 'Even as the groups became more and more alike, the remaining differences were the focus of attention' (op. cit.: 343). In many cases, differences were maintained through segregated labour markets, but as these were eroded by equal opportunities policies and the free movement of labour, territorialism gathered new strength (op. cit.: 346). In Northern Ireland, for example, the promotion of equality in the workplace has been accompanied by increased residential segregation. Across Europe different interpretations of nationhood exist within the same state territory, and there are many regions where nationalists lay claim to a larger area than that inhabited by their respective ethnic group. Where such a claim is resisted by another, dominant group sharing the same space, a satisfactory settlement often appears impossible.

In these allegedly postmodern times, we hear much about the deterritorialisation of identities through increasing interdependence, globalisation and migration as well as the much-discussed end of the nation state. But as Orvar Löfgren (1996: 166) has pointed out, instead of 'facing a future of intense deterritorialization', we may simply be failing to observe 'the different ways in which people and identities take place on new arenas and in novel forms'.

Observers of Northern Ireland often represent what is happening there as a relic from a dark European past. This perspective may be rather misdirected (Kockel 1994). It is not good enough simply to label what has been going on in Northern Ireland, the Kosovo, the Basque country and other parts of Europe as atavistic. It may be, but what is the cultural background against which we judge the events and their perpetrators? And what do such apparent atavisms tell us about our very own modernity in which they rear their ugly heads?

I am not talking here about the big acts of politically motivated terrorism but about everyday expressions of conflict in Drumcree, Ardoyne and all those other contested places and spaces, which offer to the ethnologist a possible window into the future of more than just Northern Ireland. We are witnessing there the slow, painful negotiation of nationality, citizenship and identity in their territorial context as part of a deliberate political process involving an entire regional society at the level of everyday experience. The success or failure of this process will have significant implications beyond the territorial limits of that particular conflict, wherever ethnic frontiers have been emerging in Europe. The outcome is hard to predict, but I share Köstlin's (1999) concern that our academic rationalisations cannot gloss over the increasing likelihood that, in the light of an ever-growing fear of globalisation, the real 'career' of ethnographic knowledge is only just beginning.

In this chapter, I am primarily considering aspects of the ethnic frontier in the Northern Ireland context, by re-examining three case studies (Kockel 1999b, 2001b, 2005a) that marked the beginning of my thinking about 'frontier' as an analytical concept (Kockel 1999a, 2005b). These case studies originally grew out of an earlier project examining likely impacts of the creation of a Single European Market (SEM) on regions with a contested border. I look at them now with the benefit of hindsight, but since 2005 also as a local resident, and therefore from a somewhat different perspective.

Reflecting on Europe at Drumcree parish church

In the second half of the 1990s the world came to know the name of a small country church outside Portadown, in Northern Ireland.[1] The name 'Drumcree' epitomised a whole complex of political and social tensions surrounding, primarily, different perceptions of nationality, identity and citizenship in that part of the world. Every July, the local Orange Order stages a parade to and from Drumcree parish church as part of the so-called marching season. The high point of this 'season' is on 12 July, when marchers commemorate the victory of King William of Orange over the deposed King James II at the Battle of the Boyne in 1690. Unionist protagonists see in these parades a celebration of Northern Irish Protestant culture and history, and a popular affirmation of the constitutional union between Great Britain and Northern Ireland. To their nationalist opponents, the parades are triumphalist displays reflecting the supremacist attitude of the unionists.

Chronology of a dispute

The issue of Orange parades has long been a source of controversy in Northern Ireland. In the late 1990s, despite the ceasefires by paramilitary groups, open conflict and rioting in connection with the parades escalated. Some observers suggested that the parades issue replaced paramilitary violence as a focal point for political confrontation, and although the situation in Northern Ireland has become somewhat less tense since the 1998 Good Friday Agreement (GFA), parades continue to be a thorny issue.

On 9 July 1995, the Royal Ulster Constabulary (RUC) prevented an Orange Order parade from marching along Garvaghy Road on its return from an annual service at Drumcree parish church. On 10 July, attempts were made to break through the RUC barricades and sporadic rioting broke out at Drumcree and several other areas throughout Northern Ireland. Thousands of parade supporters converged on Portadown as Orange Order leaders and senior police officers tried to negotiate a solution to the crisis. In the evening, a rally at the church was addressed by the leader of the Democratic Unionist Party, the Rev. Ian Paisley, who later, together with Ulster unionist politician David Trimble, made an unsuccessful attempt to break through the RUC cordon. On the morning of 11 July a compromise was reached with the help of an independent mediation group and the parade was allowed to proceed along Garvaghy Road, provided they did so silently. Led by David Trimble and Ian Paisley, some 500 Orangemen eventually marched down the road. Although the parade passed off peacefully, when it reached the centre of town, the two politicians raised their arms in an apparently triumphant gesture, an action that caused severe resentment among the Garvaghy Road residents.

In the light of events in 1995, the RUC chief constable decided on 6 July 1996, to re-route the parade, in accordance with Article 4 of the 1987 Public Order (NI) Order. This allows for restrictions on the route of a parade if the chief constable considers it a potential cause of public disorder. The chief constable may also, with the consent of the secretary of state for Northern Ireland, request the district council concerned to issue a prohibition order banning all parades in a particular area. On 7 July 1996, the RUC once again prevented the Orange parade from Drumcree along Garvaghy Road. In the afternoon, the grand master of the Orange Order, the Rev. Martin Smyth, declared that this time there could be no compromise. By midnight, more than 4,000 protesters had gathered at Drumcree. A four-day stand-off commenced, accompanied by disruption and rioting in many locations throughout Northern Ireland. The crowd assembled at Drumcree continued to grow, and there were increasing outbreaks of violence throughout the region. During the confrontation at the barbed wire barricades erected by the RUC and the British Army at Drumcree, three demonstrators were injured by plastic bullets. Given the scale and spread of the disorder, the resources of the RUC were stretched to their limit and additional British Army contingents were sent to Northern Ireland for support. Serious conflict erupted in Belfast, following a number of Orange parades, with several Catholic families being intimidated out of their homes. During the unrest, one person was killed and well over 100 people were injured. By the evening of 10 July, some 10,000 demonstrators had arrived at

Drumcree. On the following morning, the chief constable gave in to this pressure and allowed the parade to go ahead. Some 1,200 Orangemen marched along Garvaghy Road. Local residents had not been consulted, and rioting broke out, spreading quickly to nationalist areas elsewhere. Severe rioting in nationalist areas continued for several days. The chief constable's decision to let the parade go ahead was criticised by local community groups, politicians, church leaders and the Irish government. He defended his decision in a BBC radio interview, arguing that with an anticipated crowd of more than 60,000 Orange supporters assembling at Drumcree on the eve of 12 July, he would not have been able to enforce the ban.

As a result of the events in 1996, many nationalists in Northern Ireland lost faith in the RUC as an impartial police force. The leader of the nationalist *Sinn Féin* party, Gerry Adams, said that members of the nationalist community apparently had no rights. In a public statement on 12 July, he declared the RUC as completely unacceptable to this community and called for radical changes to its organisation. John Hume, leader of the Social Democratic and Labour Party (SDLP), and his deputy Seamus Mallon strongly condemned the government's response to the situation. The Irish government voiced its concern at the situation and expressed surprise that dialogue had not continued between the different groups involved. The Taoiseach (prime minister), John Bruton, when interviewed on television for BBC One's main evening news, severely criticised the British government for yielding to pressure from the Orange Order. Moreover, in a statement on 25 July, he condemned the British government for yielding to force, being inconsistent in its policy decisions and partial in its application of the law, thereby failing on key principles of democracy.

As in the previous two years, the parade on 6 July 1997 initiated widespread unrest across Northern Ireland. No agreement had been reached between the local Orange Order and the Garvaghy Road residents. In this situation, it fell to the new secretary of state for Northern Ireland, Mo Mowlam, to decide whether the parade should be allowed to proceed. During the period leading up to the event, the British Army presence in Northern Ireland was reinforced. RUC and British Army checkpoints were set up in the Portadown area. Women from the Garvaghy Road district formed a peace camp in tents pitched by the side of the road. A paramilitary organisation, the Loyalist Volunteer Force (LVF), threatened to kill Catholics if the parade was prevented from proceeding along Garvaghy Road. The Orange Order rejected a proposal to waive its right to march as a gesture of reconciliation. At half past three in the morning of 6 July, soldiers and police officers sealed off the Catholic housing estates along Garvaghy Road. Residents were prevented from attending mass at the local Catholic church. Instead, mass was celebrated in the open, in front of the police and army lines. At lunchtime, the parade was allowed along Garvaghy Road. Once it had passed and the security forces began to withdraw, rioting erupted once more, spreading rapidly to other parts of Northern Ireland.

In the aftermath of these events, a parades commission was established and given responsibility for decisions on contentious parades. On 29 June 1998, this commission announced that the parade to and from Drumcree, planned for 5 July, was to be re-routed. The Orange Order, in response, announced that it would march its 'traditional' route and, if prevented from doing so, would stay on the spot until allowed to proceed, however long this might take. As in the previous year, the security forces in the area were increased in preparation for the anticipated confrontation. Soldiers erected a barricade on the road between Drumcree parish church and Garvaghy Road. A trench lined with barbed wire was dug through the fields on either side. After their annual service at Drumcree on 5 July, the Orange Order first marched up to this barricade, but then returned to the church where it set up an encampment. The Orange Order announced that its members would remain encamped at Drumcree until they were allowed to march back to the town along their 'traditional' route. The grand master and other leading figures joined the protest during the day, and during the following night there was widespread rioting in Protestant areas across Northern Ireland. Between 4 and 14 July 1998, the RUC recorded a total of 2561 public order incidents. In over 600 attacks on members of the security forces, 76 police officers were injured. Most of the private houses that were damaged were owned or occupied by Catholics, and many families were forced from their homes. A number of Catholic schools were vandalised and set on fire. It was generally expected that the crisis would intensify over the weekend of 12 July, the high point of the 'marching season'. In the early hours of Sunday, 12 July 1998, three Catholic boys were burnt to death in an arson attack on their home. Although this tragedy provoked condemnation from across the political spectrum, and the escalation of the crisis was stalled somewhat, the local Orange Order in Portadown voted unanimously to continue their stand-off, and a token presence at Drumcree has been maintained by Orange supporters since then.

Over the following months, the British government engaged in long-drawn-out discussions with the two parties involved in the dispute, but no agreement was reached. There was further loyalist violence, including the murder, by a paramilitary group called the Red Hand Defenders, of solicitor Rosemary Nelson, a legal adviser to the Garvaghy Road residents. As the peace process initiated by the GFA evolved, Drumcree has become comparatively quiet. Support for the campaign of the Portadown Orange Lodges has declined and there has been much less violence. However, as recently as September 2009, the Portadown District Loyal Orange Lodge No. 1 reported on its website that 'Drumcree Protest continues each and every Sunday at 1p.m.' (www.portadowndistrictlolno1. co.uk – accessed 14 October 2009).

Territory and identity

Parades and processions going from A to B, and returning by a different route, are a common, perfectly normal and widely accepted human practice. The annual procession to and from Drumcree parish church, however, has been giving rise to major grievances on both sides because the return path leads the marching group through the *territory* of another group, who object to this. Both groups claim their civil rights and liberties, qualities of their citizenship, to defend their respective point of view. The main cause of the conflict lies in differential assertions of nationality. And yet both sides are united in a common identity not merely by virtue of their residence in Northern Ireland but through the vital role that this antagonism itself plays for their ethnic self-ascription, which forms the centre and focus of their identity. Although the general situation in Northern Ireland has improved somewhat since the GFA, during the 'marching season' in particular the two sides remain reluctant to engage in direct dialogue. What reporters (and even many politicians) have not fully recognised over the years is that the situation itself constitutes a dialogue, similar to a more or less silent game of chess, enacted in the streets. The dialogue is about territory, and about ownership thereof. By marching along Garvaghy Road, Ulster unionists claim this area as theirs. By opposing their march, Irish nationalists deny that claim.

Postmodern utopias notwithstanding, the challenge of territory still lies at the root not just of ethnic conflict but of ethnicity itself, which can be understood as 'a product of dissociation between territory and culture' (Oommen 1994: 191). Orvar Löfgren (1996: 165) suggested that '[w]e need to reflect upon what kinds of contributions European ethnologists can make to the heated interdisciplinary debate on identities and territories'. What could, or should, be the most obvious concerns of European ethnologists 'in relation to the Europe we see today, and in relation to what those disciplines closest to us study' (Christiansen 1996: 137)? Anthropologists and European ethnologists can offer their knowledge of 'culture' and their ethnographic expertise (Thompson 1997: 786). They can challenge the narrow instrumentalist notion of culture, apparent in European and national policy documents, whereby 'cultural diversity seems to be fading into nothing else than the diversity of fauna and flora' (Krasnodebski 1994: 50–1). Such instrumentalist notions prevail, and to a certain extent underpin ideas of 'unity in diversity' or a 'Europe of the Regions'; however, the hierarchy of culture over nature, implied in Krasnodebski's assessment, would not be tenable within an ecologically grounded perspective on culture, as I have outlined it elsewhere (Kockel 2002a).

The relationship between nations, states and ethnies can be reflected at both a theoretical and an empirical level. Northern Ireland, with its colourful mix of identities, is a prime example of how '[h]ybridisation and the implementation of identities ... draw attention to the crisis of the nation state and ... challenge its homogenising logic' (Caglar 1997: 177). Also, '[t]ransnational migrants bring into question the state's ability to define "the people"' (Jacobson 1996: 4); and, ironically, 'nationals of member states who exercise their right to work and live in another member state thereby disenfranchise themselves' (Brewin 1997: 235). Territory still matters, and plays a decisive role in determining belonging, both in a cultural and in a legal sense. Moreover, '[c]ultural boundaries ... may not be coterminous with identities at a variety of hierarchical levels', and we need to take into account 'the possibility that the boundaries of cultural meanings and the historical flow of events and concrete social interactions which give rise to them are not accurately definable by reference to fixed time, space and population co-ordinates' (Handwerker 1997: 806). In this context, the debate on multiculturalism 'highlights the tension between equality and difference', the need to reconcile 'the right to be different with the right to be equal' and to define any limits to such rights (Caglar 1997: 178; see also Nic Craith 2004a).

Historical and contemporary attempts to construct 'Europe' as a unity have been characterised as 'highly partial histories, as well as being extremely narrative' (Christiansen 1996: 139). Such constructions are employed to justify inclusion and exclusion. Interestingly, it appears that for the EU, inclusion is determined on economic and exclusion on cultural grounds. All this raises some serious questions with regard to European integration, and to which European ethnology may be able to offer, if not answers, at least valuable insights.

Nationality, citizenship and identity

For the purpose of this analysis, three key terms – 'nationality', 'citizenship' and 'identity' – need to be clearly differentiated. *Nationality* is understood here as attachment to a specific *community* imagined as 'natural' and defined in terms of a cultural, historically rooted nation. *Citizenship*, by contrast, denotes attachment to a specific *polity* constructed as a state in terms of rights and obligations. *Identity* is seen, on the one hand, as the totality of relationships defining an individual vis-à-vis other individuals, engendering a sense of belonging (or alienation). On the other hand, it signifies a set of markers shared by a particular group, indicating its members' attachment to a nation (nationality) or state (citizenship).

Figure 2.1 represents what seems to be the aspiration of the EU as expressed in the Maastricht and Amsterdam treaties. Nationality forms an essential part of identity, but is also an integral part of citizenship – like Scots or Welsh nationality, which are assumed to be fully contained within British citizenship. It also acknowledges that an individual's identity is larger than her/his citizenship and suggests that citizenship, like nationality, is an essential part of a person's identity.

Figure 2.2 gives a more realistic representation of the three categories in relation to each other. Most non-migrant people would probably see parts of their nationality and citizenship as irrelevant to, and even outside of, their individual identity and for many of them, citizenship and nationality, although connected (and connected also *within* their identity), would not be necessarily congruent.

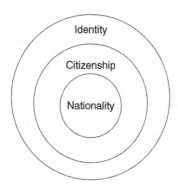

Figure 2.1   Ideal-type embedded categories

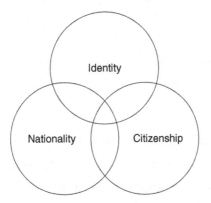

Figure 2.2 Real-type interlocking categories (settled)

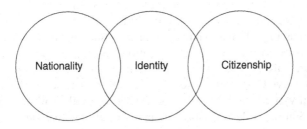

Figure 2.3 Real-type interlocking categories (migrant)

The integrated migrant population, especially in the second and later generations, is more accurately represented by Figure 2.3. Born in one territory, or at least descending from parents born there, but now citizens of another territory, this group represents the classic case of hyphenated identities, such as 'Irish-American', where 'Irish' is the nationality, 'American' the citizenship and identity provides quasi the hyphen.

Finally, in Figure 2.4 we see the alienated person. In this extreme form, few cases may fit the representation, but the graphic serves to highlight the 'end product' of the process of disenfranchisement that not just the immigrant, but especially the EU internal migrant, and even the non-migrant population might suffer in certain political and cultural circumstances. The proposal of an EU citizenship may perhaps be seen as an attempt to prevent such a situation. But is it suitable for the purpose?

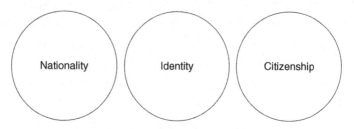

Figure 2.4 Disengaged categories (total alienation)

The EU and the idea of cultural citizenship

In the Maastricht Treaty, the EU committed itself to promoting the cultures of the member states while simultaneously respecting internal diversity *and* enhancing a common heritage. The thinking behind this project is evidently the same that informed the creation of especially the larger member states, constituting them as 'nation states' of encompassing – British, French, German – 'state-nations' that had to be constructed in the process. These nation states, presumably integrating cultural diversity across a substantial territory into 'one nation', obviously served as a model for Article 128, which states that '[t]he Community shall contribute to the flowering of the cultures of the Member States, while respecting their national and regional diversity and ... bringing the common cultural heritage to the fore'.

But how does the EU understand culture and cultural diversity? Let us look at the Commission's 1987 document (European Communities – Commission 1988: 10), *A Fresh Boost for Culture*:

The completion of the internal market implies – at the cultural level – the realisation of four major objectives ...:

(i) the free movement of cultural goods and services;
(ii) better living and working conditions for those involved in cultural activities;
(iii) the creation of new jobs in the cultural sector in association with regional development ...;
(iv) the emergence of a cultural industry which is competitive within the Community and in the world at large.

Culture consists, therefore, of goods and services provided by a class of culture professionals and administered by culture managers. It constitutes a sector of the economy and, as such, an industry where competition is a key goal ensuring smooth functioning of the market and, ultimately, customer satisfaction. Culture is, therefore, an economic activity like farming, manufacturing or banking. An anthropologist might argue, however, that all these are culturally contingent forms of socio-economic organisation or institutions. Anthropologically, 'culture' refers to the *ways* in which things are done differently in different parts of the world or by different groups – like Ulster unionists and Irish nationalists in Northern Ireland – in the same part. Therefore, culture is a process in which everyone is involved, not a product created by a few. The latter concept of culture has its roots in the historical role cultural intellectuals have played in the ideological construction of those 'national' traditions and identities that Article 128 calls 'the cultures of the Member States' – in other words, where culture has been an instrument of power politics and internal colonialism.

The Treaty of Amsterdam says little about culture. A keyword search of the Internet version only produced 'agriculture'. It speaks instead of citizenship, and also of – once again *national* – identities. The essential statements are contained in Articles 1 and 2 of the treaty. According to Article 1(5), a key objective of the EU is 'to strengthen the protection of the rights and interests of the nationals of its Member States through the introduction of a citizenship of the Union'. Article 1(8) states that the EU shall 'respect the national identities of its Member States'. Citizenship of the EU is formally established in Article 2(9), which decrees that '[e]very person holding the nationality of a Member State shall be a citizen of the Union'. This citizenship shall, however, not replace *national* citizenship. Significantly, following this treaty, citizens of the EU may now, according to Article 2(11), write to any EU institution in one of the treaty languages and 'have an answer in the same language'.

Citizenship has two principal functions. It determines who does or does not belong to 'the people' constituting a particular state. Moreover, the rules of citizenship determine the character of interactions between the individual and the state, 'the rights and obligations of the citizen, the kind of access the citizen has to the state, and the kinds of demands the state can make upon the citizen' (Jacobson 1996: 7). Consequently, 'citizenship is the linchpin of the nation-state' (Jacobson 1996). This traditional basis of the nation state is continuously eroded by transnational migration (op. cit.: 8) and '[t]he devaluation of citizenship, together with the weakening of sovereign control and the principle of national self-determination, creates questions about the legitimacy of these states' (op. cit.: 9).

Establishment of a citizenship of the EU may not be so much the emancipatory gesture as the Treaty of Amsterdam seems to present it, but rather the necessary complement of a strategy that aims to increase mobility while at the same time effectively diminishing the sovereign powers of nation states. If the depreciation of citizenship at that level is not balanced somehow, the danger is that we ultimately end up with an alienated population as indicated in Figure 2.4 earlier. Alienation breeds social discontent, signified by a lack of loyalty towards the state and its institutions, whose legitimacy is called into question. It may also be worth observing the terminology of the treaty: 'Every person holding the *nationality* of a Member State shall be a *citizen* of the Union' (emphasis added). Consequently, British citizenship, for example, comprising many indigenous and exogenous nationalities now becomes in itself, at the stroke of a pen, a nationality. The terminological ambiguity of everyday language is used here to powerful political effect. This is an issue I return to in Chapter 3.

Belonging in and to 'Europe'

Thompson distinguishes 'three categories of ethnic minorities defined in terms of their historical relationship to a dominant or majority group' (1997: 791). *Indigenous peoples* form 'relic cultures' in former colonies like the US or Australia. *Nationalities* have a potential claim to statehood, like the Scots or the Basques. Under the ambiguous label of *cultural minorities*, Thompson includes groups, such as transnational migrants, that are seeking protection of specific cultural practices and/or beliefs, but not necessarily statehood. In the 1990s, a growing awareness of the differentiated nature of cultural diversity – which is indeed much more complex than the competition of cultural producers, recognised by the EU – sparked off the debate over multiculturalism with its focus on the conflicting rights of difference and equality. This debate took different forms and directions in different political cultures as can be illustrated with reference to three major nation states in the EU: France, the UK and Germany (see also Chapter 3).

The concept of 'freedom' is, for example, interpreted quite differently in each of the three cultures (Schiffauer 1997). While in France, the essence of freedom is equality, in the UK it is the inviolability of the individual. However, both cultures converge in the way they actualise these concepts through a set of agreed rules that allow an orderly societal competition to achieve the common good through negotiation. In Germany, freedom is inextricably linked to the concept of responsibility. This presumes metaphysical knowledge of the common good to which individual interests and any mechanical equality may have to be, if not sacrificed, at least subordinated. Thus a dialectical relationship between individual and society is posited which may be described as holistic in contrast to the more particularistic relationships characterising British and French political culture. The relationship between individual and collective is an issue I return to throughout the book. Germany and France are often compared as examples of contrasting concepts of nationhood (for example, Smith 1988; Jacobson 1996) – an ethnic nation the former, a civic nation the latter. It seems to me that similarly conflicting ideas of nationality are behind the tensions we witness in Drumcree and elsewhere in Northern Ireland, where the essentially ethnic Irish nation confronts an essentially civic British nation in by now customary stand-off or open rioting (see also Nic Craith 2004a).

Ethnic resurgence, like inter-state migration, threatens the nation state (Kockel 1999a), whose elites tend to see genuine ethnic pluralism as hampering their efforts to promote democracy, and therefore as essentially anti-democratic. Over the past decades, phrases such as 'nationalism raises its ugly head again' have been used liberally, often in relation to Central and Eastern Europe. With reference to Asia, but equally applicable to Europe, Brown (1994) has identified three types of this argument. 'Ethnic chaos' theories serve to justify restrictive policies on the grounds that democracy and national unity need to be safeguarded. A second type is the postulate of a consensual *volonté general*, embodied by the state, which would be undermined by the recognition of ethnic diversity. Third, the majority principle legitimises the dominance of a cultural majority. These types are not mutually exclusive and appear together in many currently popular political arguments urging us 'Europeans' to cultivate the virtues of civic over and above ethnic nationalism. Anthropologists are understandably reserved in their support for such arguments. While acknowledging the commonsensical and humanitarian aspects of civic nationalism, they are all too aware that this same ideology has served to bring ethnically diverse peripheries under the control of culturally different centres of power, using the disguise of a common or communal ideology to 'buy off', in the first place, the local elites. In this sense, civic nationalism may be the nationalism of the imperialist whereas ethnic nationalism may be the nationalism of the colonised. German nationalism gathered strength particularly during the Napoleonic wars, and again after the Treaty of Versailles – both periods *experienced* as periods of colonisation by, first and foremost, a *civic* nation. That this ethnic nationalism turned itself imperialistic as the power of the state increased does not invalidate the basic distinction.

The EU has committed itself, in the Maastricht Treaty and, less clearly, in the Treaty of Amsterdam, to protecting regional diversity. In doing so, diversity has obviously been perceived as being of the 'cultural minorities' kind that poses no challenge to either the EU or its individual member states. Nationality has been attributed to these member states, and 'nationalities' without a state have thereby been reduced, politically and legally, to 'cultural minorities'. If their claims for the protection and promotion of specific cultural beliefs and practices are met, their 'entitlements' can be regarded as fulfilled and the pursuit of any aspirations to statehood becomes illegitimate, if not illegal. One might argue, as some nationalists in the EU do, that concessions on, for example, language maintenance are merely granted to keep movements for self-determination at bay. This may be an extreme view but the experience of multi-ethnic states, from the US to the former Soviet Union and Yugoslavia, suggests that the more ethnic identities are protected, the weaker identification with the larger whole becomes. Unless ethnic diversity in itself becomes an integral part of identity for *all* members of the state society, the 'melting-pot' ideology leads to a zero-sum game. Until Irish nationalists in Portadown join the Orange procession parading along Garvaghy Road – and are genuinely welcomed to do so – we have a problem. But once they join, the event would lose its meaning and purpose. So, once again, identity in society appears as a zero-sum game – if Irish nationalists culturally appropriated Orange marches, Ulster unionists would no longer own them in an exclusive way. Whether they would, consequently, disown them as henceforth meaningless, or grasp the nettle of multiculturalism or polyculturalism, is quite another question.

The Amsterdam Treaty, as other European documents before it, postulates hierarchical levels of belonging in the EU: the nation state and the EU. It seems that any level below the state, or, as it were, between state and union, is not considered important, merely providing additional colour to the two key levels. Yet those excluded or at best ignored sources of cultural plurality are, in fact, where people's primary identities tend to be located: as migrants, as inhabitants of cultural regions or as members of nationalities, religions or social movements. Migrant identity, nationality and citizenship, to take only one example, are problematic not only for immigrants but also for internal migrants. As a German national living in the UK for more than 25 years and teaching about Irish culture and society for much of it, I have developed a complex enough identity; having been resident outside German territory for much longer than the legal allowance of ten years, I find myself politically disenfranchised. At present, while I may pay taxes and discharge other civic duties, I do not have a vote in national elections either in my home country or in my country of residence. By exercising my right to free mobility within the EU, I have deprived myself of meaningful political franchise. Not a full citizen of anywhere, I doubt that European citizenship will make a significant difference in the short term – and suspect that many migrants will join me in this. *In* Europe we do not, technically, belong anywhere (whatever I might imagine in my mind); so, do we belong *to* Europe, at least?

The sociologist Max Haller (1994) has identified a set of prerequisites for a European identity, modelled on the 'national' identities of the nation state. These prerequisites include a territorial basis, established through history of residence and consistency of area; an integrated economy built on interdependence and free mobility; and a common culture of shared elements such as language and religion. In part, his criteria are certainly fulfilled, although they remain problematic even so. If history of residence in the territory of Europe is a criterion, then immigrants are, by definition, excluded. Consistency of area may be a suitable criterion for nation states, but a European identity is difficult to perceive on this basis, since nobody really knows where 'Europe' begins and ends, as the ongoing multidisciplinary debate on the matter demonstrates. We do indeed live in an interdependent, integrated economy but free mobility within it is available only at a price even (as my own experience shows) for socially more privileged migrants; therefore it cannot not be considered genuinely as 'free'. Common culture is the most differentiated of Haller's criteria, and he gives language and religion as key examples, although the territorial definition of a consistent area also implies more material cultural markers. With regard to language, English is indeed making great inroads and may well become the language of the EU. But is this the native tongue spoken by a minority of Europeans who live on some offshore islands in the north-western corner of the EU or is English the language of the global cultural hegemony at the beginning of the twenty-first century, the US? With regard to religion, should this criterion be used merely to exclude Europe's definitive 'Other', Islam, from our common culture, thereby denying the great cultural contribution of Islam to the formation of what might pass for a 'European' culture (if there is such a thing; cf. Nic Craith 2004b, 2007)? Or should it be applied to deny basic civil rights to either Protestants or Catholics in Portadown? Haller sees these problems, and acknowledges that a European identity cannot be formed in analogy to the national identities that have been constructed over the last few centuries.

Drumcree and European identity

Oommen (1994: 189–90) points out that the fusion of citizenship and nationality was a goal of the nation state, which sought to accommodate different nationalities within a common citizenship. Where nationalities are strongly developed as in the UK, this attempted fusion may create centrifugal forces. Common citizenship is necessary to facilitate such goals as the free mobility of labour, but the strengthening of multiple nationalities is essential to provide a 'basic anchorage for cultural diversity' if such is genuinely seen as an asset. The Free State of Bavaria, or the Autonomous Community of Catalonia, represent and safeguard the nationality of their nationals who are, at the same time, fully Bavarian and fully German (see Chapter 4), fully Catalan and fully Spanish. They can be both, because being Bavarian or Catalan is different from being German or Spanish. Why this should be so is a question that lawyers or economists would probably find difficult to answer. Sociologists, political scientists or psychologists might fare somewhat better, geographers possibly better still. By virtue of their subject, but also the history of their discipline, European ethnologists should have something more to say on this question. In these allegedly postmodern times, we hear much about the deterritorialisation of identities through increasing interdependence, globalisation and migration as well as the much-discussed 'end of the nation state'. But, as pointed out earlier, instead of 'facing a future of intense deterritorialization', we may simply be failing to observe 'the different ways in which people and identities take place on new arenas and in novel forms' (Löfgren 1996: 166).

Arguably, representations of the Northern Ireland conflict as a relic from a dark European past have been rather misdirected (Kockel 1994). What we have been witnessing in Northern Ireland since its creation as a political entity is the slow, painful negotiation of nationality, citizenship and identity in their territorial context as part of a deliberate political process involving an entire regional society at the level of everyday experience. The success or failure of this process will have significant implications well beyond the territorial limits of this particular regional case study. As a European ethnologist, I have been observing Drumcree, and all that it symbolises, as a possible window into the future of more than Northern Ireland – perhaps 'Europe', whatever that might be.

A politicised cultural landscape[2]

The conflict in Northern Ireland has been analysed in detail by historians, political scientists, sociologists and some geographers.

Anthropological and ethnological studies have primarily concentrated on the symbolic construction of (group) history and political territory, augmented in recent years by studies of (mostly Gaelic) language and identity, and studies using psychoanalytical approaches. In the context of the Irish Republic, identity discourses relating to landscape and nature have been studied for some time (e.g. Sheeran 1988a, b; Edwards 1996; Gibbons 1996), but for Northern Ireland, with the exception of Henry Glassie's (1982, 2006) work, little has been done in this field as the conflict appears to have dominated research agendas. Even cultural geographers, who should be open for these kinds of issues, have tended at times to use rather peculiar arguments to relegate any engagement with the topic from the empirical to the discursive level, where any connection with the everyday is easily lost.[3] However, some individuals and groups, especially among Protestants, have used older geographical work on related themes[4] to justify their separatism. Their concern seems to be less with a particular, geographically defined landscape, but with an ideal-type 'frontier culture' suited to winning dominion over a wild and threatening 'nature' (including the barbarian 'natives').

Whereas contemporary flashpoints, such as Drumcree, may acquire symbolic status in the everyday interaction between different groups, there has been a much broader symbolic and practical incorporation of the cultural landscape into the conflict, as the discussion in this section illustrates. Two contrasting yet connected discourses need to be carefully distinguished here. To do this, I have used the German terms *Trenngrenze* and *Mischgrenze*, borrowed from science (Kockel 2001b). *Trenngrenze*, which can be translated into English as *boundary*, is oriented more towards the 'here' and 'now', whereas *Mischgrenze*, translated into English as *frontier*, is oriented towards the 'there' and 'then'.[5] The so-called Drumlin Belt will serve here as an example for the *Trenngrenze* – a hillscape that many Northern Irish Protestants regard as a dividing line between the British North and the Irish South of the island. The *Mischgrenze* will be explored with reference to the two main ethnographic open-air museums in Northern Ireland, the Ulster Folk and Transport Museum and the Ulster-American Folk Park, with particular focus on the latter.

The discovery of 'natural' boundaries

Particular, basic structures of the physical landscape – such as hills or rivers – can be perceived as 'natural' and therefore God-given limits of specific cultural spaces. In the case of Northern Ireland, the Drumlin Belt, a hillscape made up of glacial deposits in the southern part of Ulster, is such a structure. In this landscape the combined effects of physical, religious and economic factors have kept the suspicion of an established border awake (Evans 1992: 31). The majority of inhabitants of this landscape are Protestants. In the course of the so-called plantations, that is, the organised settlement of the northern parts of Ireland by Protestant colonists from Scotland, England and Wales during the seventeenth century, the best soils had been occupied by the settlers, who turned former pasture into more lucrative arable land, while the rocky uplands between the Drumlins, as well as the swamps and bogs in the lower areas, remained mainly in Catholic hands. Evans (1992: 30) describes this settlement structure as 'Protestant islands in a Catholic sea'.

The postulate of a natural *Trenngrenze* between north and south, fortified – as imperial marches everywhere and at all times – by faithful settlers, rests on a rather selective interpretation of the much more differentiated work by Heslinga (1962) and Evans. Since the usefulness of any settled culture landscape as a symbolic line of separation is necessarily limited, the idea of such a separation is historicised by emphasis on putative evidence of fortifications dating back to pre- and early-historical periods (Figure 2.5). The gently rolling hillscape of the Drumlins is being stylised as an insurmountable mountain range, and those defensive structures are taken as proof that the island has been divided at least since the dawn of history. Earthworks such as the Dorsey can be found all over Ireland and Europe, and archaeologists tend to interpret them in a local rather than a supra-regional ethnic context. In Northern Ireland, however, they have a particular ideological significance.

While his writings cannot be discussed in detail here, the work of Ian Adamson should be noted in this context.[6] Adamson's main thesis is that, before the arrival of the Gaelic Irish, the northern parts of the island were settled by the Cruthin. Under pressure from the invading Gaels, the Cruthin barricaded themselves behind those earthworks, but finally had to give up their resistance and flee to Scotland, from where their descendants returned to their native land only in the seventeenth century. Adamson and others interpret this representation of the past as a contribution to reconciliation since it ascribes both sides in the conflict a right to be on the island. While archaeologists (e.g. Mallory and McNeill 1991) question the credibility of the Cruthin theory, its nationalistic undertone makes it highly attractive for groups on the radical Right.

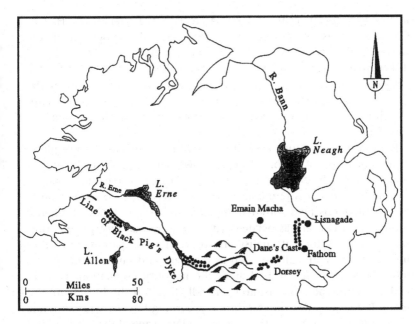

Figure 2.5 'Natural' and 'ethnic' boundaries in Ulster
Source: Map drawn by Michael Murphy, University College Cork, and reproduced from Kockel (1999a).

The settler myth

A historical comparison of physical and demographic maps of Ireland shows interesting, if not entirely surprising parallels. The less fertile mountain and bog landscapes that are dominating the western province of Connacht, together with similar areas in the southern province of Munster and western parts of Ulster, are primarily poor, peripheral and relatively underdeveloped in terms of social and economic indicators. At the same time, however, they are regarded as particularly 'Irish'. This 'Irishness' finds expression not only in the Gaelic-speaking areas along the west coast, but also in the especially photogenic, 'typically Irish' landscapes that adorn the pages of contemporary tourist brochures. The association of 'Irishness' with the wild landscapes of the West has at least some of its roots undeniably in the romanticism of the Irish nationalist movement of the nineteenth century; however, critics of 'romantic Ireland' tend to neglect the historical influence of colonisation on the creation of that imagery. In the course of what might nowadays be referred to as 'ethnic cleansing', many inhabitants of Ireland, especially in the seventeenth century, were faced with the alternative: 'Hell or Connacht' (Ellis 1988), where the latter signified resettlement on the western periphery of the island. The former often meant a further choice: death in Ireland, or deportation to the colonies in the New World. One result of this policy was the increasing overpopulation of the few moderately fertile areas on the west coast, especially Connacht, which consequently were hit particularly hard by the Great Famine of the 1840s.

For Northern Ireland, a similar ecological polarisation took place, as a comparison of historical maps shows. The Republic of Ireland of today seems to be demographically less polarised, whereas in Northern Ireland there remains a notable degree of congruence: on the one hand, peripheral landscapes, often designated as various types of natural protection areas, are associated with Catholic majority populations, strong support for nationalist and/or republican parties and other indicators of 'Irish' culture, such as language and music, while in the more prosperous agricultural areas the majority population tends to be Protestant, vote unionist and generally feel 'British'. In the mind of many Protestants this differentiation according to clearly recognisable environmental characteristics, between a landscape of the 'settlers' and a landscape of the 'natives', becomes an expression of inherent cultural differences. Where these differences cannot be read unequivocally in the shape and vegetation of the landscape itself, strategically placed symbols – mainly posters, flags and garlands, sometimes also parades, such as the one to and from Drumcree parish church – are used to reinforce this interpretation.

The use of symbols is also necessary because the sharp distinction between 'settlers' and 'natives' according to ecological criteria alone does not stand up to historical–demographic scrutiny.[7] All the same, the assumed contrast between a 'cultural landscape' of the 'settlers' and a 'natural landscape' of the 'natives' has survived as an extrapolation of stereotypes handed down almost like cherished traditions. The representation of Ireland as a wilderness beyond the civilised world, populated with barbarians, goes back to classical antiquity, and was embellished during the period of the Norman conquest, in particular by Giraldus Cambrensis, whose image of Ireland should exert a defining influence for several centuries (Dohmen 1994: 20–1).

During and after the Reformation, this characterisation of Ireland as primeval – in stark contrast with a cultivated England – was reinforced by the spread of a perception of Catholic culture as pre-civilised heathenism. Thus the settlement of the imperial marches, as they expanded ever further westward, acquired a second dimension. The military aim of fortifying the Empire in this world was extended by spiritual fortification of the Empire into the great beyond, in the sense of a mystical act of deliverance that the 'settlers', as 'culture bearers', performed on the 'natives' as wild, heathen 'children of nature'. The subordination of Ireland was thus mythically transfigured and became, closely connected with the cultivation of the wilderness per se, a symbol of Protestant dominion over nature.[8]

In the course of this mythicisation, the taking over of the most fertile parcels of land by the 'settlers' becomes the cultivation of large parts of the wilderness; the prosperity achieved on these soils becomes an indicator of the success of the civilising effort, in particular when compared to the evidently poorer rocky uplands and boggy lowlands that are worked by the 'natives'. Following ancient and medieval representations of Ireland, the entire island is imagined as consisting of rocks and bogs, from which the fruits of the soil were wrested by the hard labour that typifies only the 'settlers'.

Telling the story of the frontier

The myth of Protestant dominion over 'nature' was given a folkloristic memorial with the establishment of the Ulster-American Folk Park, located near the town of Omagh in County Tyrone. Its geographical position – on the edge between the fertile agricultural land of the county and the wilderness of the Sperrins mountain area with its bog and heather landscape – was quite accidental, but nevertheless symbolic in itself, and for a long time regarded as such by people in Northern Ireland.

In the ethnological part of the Ulster Folk and Transport Museum, following the ideas of its initiator, Estyn Evans, an attempt is made to represent the 'common ground' through a more or less synchronous cross section of material culture in Ulster around the turn of the nineteenth to the twentieth century, giving appropriate attention to different cultural traditions. This representation of an 'Ulster as it was' is directed at the cultural inner core, the essence of 'Ulster' (whatever that might be). By contrast, the Ulster-American Folk Park tells, in linear, diachronic form, a story of development. Moreover, its name signals a conceptual alignment with the 'theme park' sector of the cultural industries, fashionable at the time of its establishment, rather than the comparatively old-fashioned, traditional open-air museum.[9] The focus is on 'Ulster as it developed', more precisely: on the human inhabitants of Ulster and what became of them. In the exemplary presentation

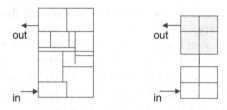

Figure 2.6 Schematic representation of different paths through the Ulster Folk and Transport Museum (ethnographic part) and the Ulster-American Folk Park
Source: Adapted from Brett 1996: 106.

of one particular Protestant family, attention is drawn to the inner core, the essential factor that made this development possible – the Protestant utopia.

That story is structured by the very design of the museum. At the Ulster Folk and Transport Museum, despite geographic–thematic clustering of the buildings, there is no predetermined route of progression through the various elements of the museum's story, no singular, specifically intended text. The Ulster-American Folk Park achieves such an ordering at several levels at once. In a case study, David Brett (1996: 106) highlights the 'strait-gate of the emigrant ship', the central display that connects the two parts of the museum – 'Ulster' and 'America' (Figure 2.6). Brett compares the passage through this central display to a rite of initiation. Reinforcing the experience is a second level of ordering. Brett observed that visitors may move freely around either part of the museum, but that the transition from one to the other, it seems, '(like emigration?) is irreversible' (Brett 1996: 106). However, that freedom of movement required a thorough knowledge of the museum grounds. In the 'Ulster' section, the museum resembles a thick forest through which there are few signposts, not very prominently displayed. In 'America', by contrast, the path suddenly becomes clear, open and free, including open vistas of the surrounding landscape.

The course through the Folk Park begins in the exhibition building, where the theme of emigration is unfolded through the skilful combination of display boards, material exhibits and reconstructed scenes from the migrant experience. The migration process is by no means glorified or even only sanitised; the display highlights the difficult conditions during the journey and after arrival in the new land. At the end of the exhibition, a gently rising path leads into the open, or rather, into a thick deciduous forest indicating the natural landscape of the region in the seventeenth and eighteenth centuries. Even on a bright summer day, the path now leads the visitor into a semi-dark environment from which a little further on the contours of a single-room cottage emerge, the kind of cabin that would have been typical for the Sperrin upland areas during that period. At this point some visitors lose direction in the thicket before rejoining the path leading them to the next building, a blacksmith's forge. Next in line is the dwelling of a weaver and gradually the primeval forest recedes, the path becomes more easily discernible. The houses become larger and their inhabitants evidently more prosperous.

On the edge of the forest lies the restored farm of the Mellon family, who emigrated to America in 1818; their descendants donated the land and financial resources to establish the Folk Park, and the museum tells the family's story of migration as an example. An artificial viewpoint, located behind the farmhouse, offers the first open vista in the 'Ulster' section, although only in one direction. Beyond the forest that surrounds them, the visitors see, nearby, orderly fields with the almost menacing, dark bulk of the wild Sperrins beyond – a message that could hardly be more graphic.

After returning to the still fairly dense forest, which is now increasingly interspersed with cultivated fields, the visitor passes more rural buildings before entering a small town high street, which features a shop and a pub. At the other end of the street there is a large gate through which the visitor reaches a dimly lit harbour scene with a sailing ship lying at the quay. The atmosphere has been created with a remarkable sense of eeriness. The visitors have to descend into the belly of the inhospitable sailing ship in order to proceed to the second part of the museum – 'America'.

Here the houses are big, beautiful and clean, their lines are straight. There are proper roads. After leaving the town of arrival in the New World, the visitor reaches a charming farmstead built on a clearing, well fenced, and is greeted by the smell of baking from the log cabin. Even here, in the countryside, this 'old America' smells surprisingly familiar, quasi-homely and not as alien as parts of the 'old Ulster'. Through these subtly engineered sensual experiences the impression of a successful colonisation is connected positively with Protestant culture in today's Northern Ireland. The early Presbyterian perception of Ireland as a 'new-found land' (Brett 1996: 114–15), in need of being cleared and subjected to rational methods of cultivation, comes to the fore here. In a play of metaphors, the 'Ulster' that is being left behind stands for the 'Ireland' in which the 'settlers' arrived earlier, just like today's visitors in the recreated primeval forest. The 'America' reached by symbolic naval passage stands for the 'Ulster' that the 'settlers' have – or at least would have liked to have – made of Ireland. Through sacrifices and hard work, wild nature has finally been subdued.

Shortly after the first log cabin, the visitor arrives at the splendid new estate of the Mellon family, whose tracks this journey through time has broadly followed. Here, in the neatly geometrical order of the New World, the goal of the journey has been reached. The forms and layout of the estate remind the observer of the old advertising slogan of the southern German chocolate manufacturer Ritter: *quadratisch, praktisch, gut* (square, practical, good). While Brett's (1996: 116) assessment of the ensemble – as the settlers' mythico-religious ideas of the 'Heavenly City' or the temple itself, translated into architecture – may seem a little far-fetched, it is entirely comprehensible. A more detailed investigation of the relevant sources than can be attempted here would be needed to establish whether the Ulster-American Folk Park was indeed, as Brett argued in his case study, conceived as a philosophical garden for a 'Protestant tradition' that has been modelled in its layout and narrative on the symbolism in John Bunyans *Pilgrim's Progress*.[10] But even without a distinctly theological frame of reference it is possible to deduce, from the epic of mastery over 'nature' outlined here, some interesting observations with regard to Protestant identity that cast light on everyday cultural and political life in the region. It is not important for this purpose whether the story of how (Catholic) 'nature' has been overcome by (Protestant) 'culture' has indeed been consciously, purposefully and skilfully stage-managed in every detail as it is retold here. What matters more is that this is precisely the story visitors – with inevitable personal variations – have experienced.

Reflections

Any religious connotations apart, the symbolic ensemble of the Ulster-American Folk Park also suggests an ideology of a kind of 'internal emigration' in Northern Ireland and the wider region of Ulster. The metaphorically well-ordered 'America' literally turns its back on the indomitable land of Ulster/Ireland. It is not insignificant that the clearing of the new Mellon farmstead opens out to the northeast and offers a view of the 'wild Sperrins'. Just as with the viewpoint at the Mellon farmhouse in 'Ulster', this upland area lies *behind* the house. And yet, the way the path is laid ensures that it remains in the visitors' mind and, as they enter the farmstead in the New World, is present as that which – really as well as metaphorically – now lies behind them.

Leaving the farmstead, the path returns to the exhibition building past the estate's geometrically laid-out herb garden, the Sperrins looming in the background. If the clearing opened in the opposite direction, the visitors would see orderly fields enclosed by hedgerows. As it is, the path leads back to the primeval forest, preparing the visitors for their return to the 'wild Ulster' of their own time.

The migration epic is deeply rooted in the ethnic self-perception of the Ulster-Scots (see, e.g. Fitzpatrick 1989; Dawe and Foster 1991; Nic Craith 2001) – that group in Northern Ireland which traces its ethnic origin to Scottish settlers, and often beyond to the Cruthin. The private story of the Mellon family is told in the Ulster-American Folk Park as a parable for the same kind of settler spirit, related through an attractive and varied educational programme. 'Ulster' and 'America' are equated as paradigmatic frontiers par excellence, where destiny accords 'God's frontiersmen' (Fitzpatrick 1989), the Protestants of Ulster, an active role in God's continuing, everyday creation of the world.

From these two fundamentally contradictory perspectives, the problems with bringing the conflict to a close can be interpreted hermeneutically. On the one hand, there is the internal exile from a land experienced as wild and barbaric but from which there is no escape. This is the secular aspect of the story told in the Folk Park – even in the New World, the old wilderness is present in the background. On the other hand, there is the awareness of a divine mission to improve a world that does not want to be improved. As foundation for identity, both positions are contradictory; at the same time, they depend on each other in an almost tragic sense. This identity, founded on the process of 'frontiering' that is experienced as vocation to contribute to the divine work of creation, would lose its raison d'être with the peaceful resolution of the conflict, which ultimately would leave no political room for such a process. In order to maintain their identity, Ulster Protestants, like the hero in J. F. Cooper's *Leatherstocking Tales*, are compelled to move on with the frontier (cf. B. Schmidt 1994). Where that is not possible, the frontier as such, and the awareness of its existence in the everyday, must be retained. In this sense the Folk Park's conscious celebration of Ulster's American heritage arguably fulfils a memorialising function that goes beyond educational historiography – it emphasises a heritage of the future, the hope of escape from the drudgery of an 'old' world to a 'new' one.

However, over the past few decades there has been a re-evaluation of 'wilderness' and the Protestants of Ulster have not remained unaffected by this. What Belinda Loftus (1994) could still describe as two contrary perspectives on Northern Ireland has effectively merged into a single one, and that perhaps for quite some time already. Insofar as generalisations are permissible here, it may be noted that Catholics in Ireland are drawing their identity in no small part from a perception of 'wild' places (Sheeran 1988a, b) and landscapes, while Protestants define themselves more via ideas of geometric order and the perception of movement and mobility as 'progress'. In that sense, Loftus still has a point and these different frames of reference for identity explain, albeit only to a certain extent, the differences of perspective between the two groups with regard to migration: Catholics have regarded emigration as an injustice inflicted by the colonial power, while Protestants have regarded it more as the fulfilment of divine providence. These stereotypes are mutually maintained by both sides as a coherent pattern of distinction between 'own' and 'other', regardless of how individuals feel about, in this instance, their migration experience.

Already before the GFA of 1998, a convergence of imagery had become evident, most notably in tourist advertising. Northern Ireland, which had long marketed itself as a tourist destination through what were essentially 'English' images of tea parties in the park, golf and classicistic architecture, has in recent years increasingly located itself, in terms of touristic representations, as part of the 'Celtic Fringe'. Moreover, within that imagery it projects itself at a juncture where it is increasingly acknowledged as having been located historically and culturally as well as geographically: between Ireland and Scotland. Thus Northern Ireland is now being re-evaluated, in a positive sense, as part of a larger 'wilderness' that needs to be sustained as a resource for development. That a new peripheralisation of Northern Ireland is implicitly contained within that re-evaluation is just one of the contradictions of this process, contradictions that could ultimately lead to an intensification of culture conflict in the region.

On the other hand, the cultural re-evaluation of landscape and nature discourses holds the potential for a reappraisal of the conflict. It is not only the two museums mentioned here that have moved significantly in the direction of so-called shared narratives over recent years. Within the museum sector, projects include the Tower Museum in Derry/Londonderry and the Navan Fort Heritage Centre near Armagh. The Tower Museum attempts to re-present the controversial history of the city in ways that both sides can identify with; leading members of both sides have been involved in the project. Navan Fort, once the political centre of the prehistoric province of Ulster, located at the northern edge of the *Trenngrenze* represented by the Drumlin Belt, tells the story of ethnic origins in a way that seeks to do justice to both versions and interprets the ancient myths as part of a common cultural heritage. The Ulster Folk and Transport Museum and the Ulster-American Folk Park are pursuing similar goals. These developments are to some extent the result of public subsidies for initiatives aimed at improving cultural relations within Northern Ireland. But such 'shared narratives' alone will not be enough to turn into reality the dream that Estyn Evans was trying to give expression to with the foundation of an Ulster folk museum – the dream of what might be called a *'Heimat* Ulster' that would be jointly owned by all those who live in this part of the island. The concept of *Heimat* will be explored further in later chapters.

It seems to me that, as indispensable foundation for such a project, we need to develop a deep understanding of historical cultural connections, contexts and identity formations in the everyday life of the region. But anyone prepared to become involved in this should take note of the observation that, after several years of a 'peace process' that has been reasonably successful almost in spite of its protagonists, the topic has lost little of its old explosive force in the region. As I am writing these lines, news of another car bomb in Belfast is coming through.

Ambivalent location: 'Stroke City'

Despite several decades of 'deterritorialisation' resulting from 'globalisation', reference to place remains an important means of spatial orientation. Through place references an individual's associations with societal events and movements can be located historically in the form of personal–biographical relationships. In the course of increasing substitution of a *genius fabulae* (sense of story) for a *genius loci* (sense of place), places are being used in new ways as anchors for identity.

In its constructivist extreme, this is a reflection of late postmodernity's 'anything goes' approach to identity formation, which maintains that we can appropriate any place as we please as long as we can somehow prop our individual stories over them. However, there are deep links between story and place that have little to do with the superficiality of the postmodern identity warehouse with its neoliberalist identikits. Some of these links will be explored later in the book. At this point, and in the spirit – often attributed to European ethnology – of seeking to understand Europe from its periphery, I want to move on from the rural settings to look at an urban place on the periphery of Europe, remaining in the island of Ireland. Having moved westward, from Drumcree along the Drumlin Belt to Omagh, we arrive in the north-west of Northern Ireland (which, as the locals never tire of pointing out to visitors, is located to the south of the far north of 'Southern Ireland'), in Derry/Londonderry. In this city, distinctive versions of identity are crafted in the interplay with a respective 'other'. We are dealing here with two places in the same physical-geographical location, where attempts have been made with variable success to achieve a shared historiography of the city's contrary cultural heritage.[11] Similar to that city on the other end of Europe, which 'speaks of Jewish *Vilne*, Polish *Wilno*, Russian and French *Vilna*, German *Wilna*, Byelorussian *Vilno* and Lithuanian *Vilnius*', its 'different topological realms might share the same terrain, but they lead to strikingly different experiences and memories of the place' (Briedis 2008: 14).

The popular reference to 'Stroke City' designates a number of different constellations of place: In 1689, the Rev. George Walker's *A True Account of the Siege of London-Derry* hyphenates the city's name, indicating the connection between an older settlement and the London guilds that parcelled out the frontier during the plantation of Ulster, the early period represented in the Ulster American Folk Park. Later, especially during the period from about 1968 to 1998, locally referred to as 'the Troubles', the use of one or other designation – 'Derry' or 'Londonderry', for Gaelic speakers also 'Doire' – became a badge of political identity. Generally speaking, the use of 'Derry' is associated with Irish nationalist and republican identities, the use of 'Londonderry' with Ulster unionist and loyalist ones, although in the late 1980s a local unionist representative told me in an interview that nobody would use 'Londonderry' locally unless they were making a political point. Nowadays, various compromises are in use, especially since the GFA of 1998, such as the spelling 'L'derry' on road signs, the alternating LCD displays on the local buses, or the convention that BBC Ulster will employ both designations according to a carefully balanced rule.

Dating back to a monastic foundation by St Columba in the sixth century, known under the name 'Doire Colm Cille' (oak grove of St Columba), the modern city was founded during the plantations, the settlement of British colonists, in the early seventeenth century. At that time, city and county were granted to the London guilds, a connection that found expression in the addition of the tag 'London-'. Irish nationalists therefore sometimes refer to the colonial city as 'London's Derry'. Dating back to the same period is the symbol of the city, its almost completely intact walls, which are celebrated by all sides locally as Europe's last medieval fortification.

The walled old town lies on a hill on the western bank of the river Foyle, where the river expands to form a kind of lagoon. In popular parlance the western bank is therefore called 'the Cityside'. Since the partition of the island in 1925, the municipal boundary here has also become a state border. As a result, the settlement structure has expanded primarily along the eastern bank of the river, popularly referred to as 'the Waterside'. However, this expansion has also skewed the social and ethnic profile of the city. In the course of 'the Troubles' since the late 1960s, the majority of those inhabitants of the Cityside who feel allegiance to the Union with Britain, belong to one or another of the various Protestant denominations, and predominantly define themselves ethnically as 'British', have moved to the Waterside or further eastward. Only in the fountain area, on the southern edge of the walled city just outside Bishop's Gate, an 'ethnic enclave' remains whose inhabitants describe themselves as 'Londonderry Westbank Loyalists' who are 'still under siege'. The 'siege' metaphor refers to the 1689 Siege of Derry, a key event in the identity discourse of Ulster Protestants, while the choice of 'Westbank' evokes fully intended associations with Israeli settlers in the Palestinian territory west of the river Jordan. For the other inhabitants remaining on the Cityside, the city is commonly known as 'Derry'. That also appears to be the predominant vernacular usage, regardless of ethnic association. A difference appears to exist mainly with regard to the usage in printed or otherwise 'public' formats vis-à-vis private conversations. Local newspapers carry their geographical location in their banner – the *Derry Journal*, based on the Cityside, compared to the *Londonderry Sentinel* from the Waterside – and various, often unwritten, conventions such as the BBC's practice referred to earlier are in operation for public purposes, ceremonial or otherwise. Some years ago, the city's mayor managed for his entire year in office not to mention the name of the city once, using instead a range of common euphemisms, including 'this city by the Foyle', 'our Walled City', or simply 'this beautiful town'. In private conversations, however, the use of 'Derry' is ubiquitous on both sides of the cultural divide. Even two major events in the Protestant calendar, linked to the annual commemorations of the 'Siege of Derry', are led by the 'Apprentice Boys of Derry'; few see any contradiction in this use of the city's name.

The debate about the name has been going on for years and did not start with the re-naming of the city council as 'Derry City Council' towards the end of the previous century. In 2003 the city council undertook initial steps to change the city's official name to 'Derry'. Because cities are given their status by Royal Charter, any change of name requires the consent of the monarch. Among the inhabitants of the Waterside, this move by the city council caused both anger and resignation.

Protestants,[12] who are in a minority in the city but in a majority on Northern Ireland as a whole, have expressed their preference for the retention of 'Londonderry', whereas the majority of the city's inhabitants appear to prefer 'Derry'– leaving the Queen and her prime minister in a quandary. As so often, commerce has overtaken political progress (or become fed up with the lack of it): tourists in the city have long been able to purchase identical postcards imprinted with the different names.

Not only its name but also the location and historical geography of the city are heavily symbolic. The oak grove from which the city derived its original name has long since disappeared. Below it stretched a swamp area, drained and developed as the modern city grew, called 'the Bogside', where the indigenous Irish working class lived. This and the area on the hill beyond, called 'the Creggan', were strongholds of the Irish Republican Army (IRA) during 'the Troubles'. Rising above this swamp settled by the 'natives' (to continue the narrative discussed in the previous section) was the massively fortified 'city on the hill', reassurance for the 'settlers' and a permanent menace to the 'natives'. Down in the Bogside, among the council flats and terraces, Palestinian flags are counterpointing the Israeli associations in the Fountain and on the Waterside – powerful symbols in a local discourse that hardly need further explanation.

As almost everywhere in Northern Ireland, painted kerb stones and garlands in the colours of the respective nation of association signal the ethnic self-ascription of streets and blocks of houses in the city – blue, white and red for Britain; green, white and orange[13] for Ireland. Here there are also a range of other sites of memory. The best known are in the Bogside. These include the street corner where most of the victims of 'Bloody Sunday' died in 1972[14] and the well-preserved gable end of one of the old houses, with the slogan: 'You are now entering Free Derry' painted on it – a reference to 1969, when the slogan was first painted to signal that the Bogside had become for some time a 'no-go' area for the Royal Ulster Constabulary, following unrest sparked by the ambush of a civil rights march earlier that year. It may just be a convenient coincidence that this preserved gable is in full view of the Apprentice Boys of Derry's assembly hall and the platform on the city walls where Protestant marchers gather at certain points of the ceremonial year.

One of these occasions is the burning of a gigantic effigy of 'Lundy the traitor'. Lundy was commander of the city at the time when the troops of King James II laid siege to it in 1689. In December 1688, when James sent new troops to be garrisoned in the city, the gates were closed to them – according to legend by 13 apprentice boys. This event is commemorated by the Apprentice Boys of Derry club, an organisation founded to keep the memory of events around the siege alive. The club parades on certain key dates throughout the year in a style similar to that of the Orange Order.[15] Every year its members build an effigy of Lundy, several metres tall, which is then ceremonially burnt at the end of a parade that is held on the first weekend in December. According to a tableau in the Apprentice Boys hall, this ceremony is intended to honour the independent individual who acts responsibly – the kind of person that Lundy, according to that same interpretation, was evidently not. Another festive occasion is the annual celebration of 'the relief of Londonderry' which ended the longest siege in British history. According to the club's currently projected self-image, the sole purpose of these pageants is the preservation of tradition that promotes tourism, and thus creates or at least secures jobs. Over recent years, with financial support linked to the peace process initiated by the GFA, a multi-day festival has grown around this parade. Under the title 'Maiden City Festival' – a reference to the fact that the city's defences were never breached – this series of events is presented as an attempt to build cultural bridges between the two cities in this place. Whether the choice of title was the most auspicious for the purpose, considering that it clearly refers to the historical preferences of one side in the conflict, remains open to debate.

At least in the old town and in the relationship between the club and the city, old battle lines appear to begin to crumble. For example, leading members of the club have been actively involved in the establishment of the Tower Museum, mentioned in the previous section. This museum, set up in a reconstructed tower house inside the city walls, attempts a skilful – and largely successful – interpretation of the city's history that enables both sides to identify with the narrative. Moreover, an almost playful use of symbolic citations can be observed lately. One example of this is the Tower Hotel. Like the Tower Museum, to which its name refers, it is situated inside the walls; it overlooks the Bogside and its own tower vies for prominence with that of the Apprentice Boys hall, which is located higher up on the hill. As if that were not enough, the name of the hotel, written on the eastern side of the tower, could practically only be read from one street within the old town, and from the Waterside: Tower Hotel Derry. Anyone who knows that until a few years ago the Catholic majority in the city was banned from even walking on the walls will understand the sting in such a harmless architectural ensemble. Here territories are being reclaimed in an almost postmodern way. At the same time, spaces are created in which new encounters are becoming possible. The intercultural programme 'Cultures of Ulster', which celebrated the newly discovered[16] Ulster-Scots culture along with Irish-Gaelic and English-British cultural expressions, was launched with a gala dinner at the Tower Hotel. However, it appears that in the course of recent renovations, the sensitive reference to the city name was removed from the eastern face of the tower.

Considering the symbolic weight of the city walls, it is hardly surprising that local discourses of identity and tradition revolve significantly around these walls and the location of various places and events in relation to them. Alongside this major element of the built environment, murals are another important feature. Belfast has become quite famous for its varied murals, but Derry/Londonderry also has some interesting examples (Guildhall Press 2008), and there is even a locality-specific style, developed by a group collectively known as 'the Bogside Artists'. When in the course of urban renewal the old houses in the Bogside were replaced with blocks of flats, this created enormous blank walls that virtually cried out for painting. Over the years a public open-air gallery was created. The murals form 'the people's gallery' (Joseph 2001) of public art, created explicitly as a contribution towards healing and reconciliation, showing scenes from 'the Troubles' and some more recent images of the peace process.

Murals can be found at a number of locations across the city. The murals of Derry vary greatly in style and content, from aggressive to humorous, but generally express the newly found self-assuredness of a culture on the way forward. The murals of Londonderry, by contrast, speak more of a culture in retreat. That applies not only to the two murals in the fountain that commemorate the battles of 1689/90, but also to the murals on the Waterside, which cover a wider range of themes. The siege is also a topic here, including some murals where the city is unproblematically called 'Derry'. There are cynical citations of 'the other side's' symbolic repertoire, for example in a mural based on a famous vinyl album cover by Iron Maiden, which shows a uniformed zombie-like character climbing over ruins, identified as the Bogside by a detailed sketch of the 'Free Derry' gable end and a mural by the Bogside Artists visible beyond it. Other representations include the celebration of the aforementioned Ulster-Scots culture and the frontier epic discussed earlier, especially its American dimension. The Ulster-Scots who migrated to North America, particularly in the eighteenth century, are portrayed here as the founding fathers – quasi the cultural backbone – of the US. From the perspective of Irish nationalist critics, this is seen as a case of the colonisers appropriating the discourse of the colonised for themselves, albeit as heroic men (mainly) of the 'frontier' rather than as emigrants due to need.

Just as the Irish nationalist master narrative of colonisation followed by transatlantic migration has been appropriated by 'the other side' for some time, so the master narrative of the endangered language has been discovered more recently (Nic Craith 2001). Since the middle of the 1990s, a new cultural industry has sprung up promoting Ulster-Scots language and folklore. Thus the Irish nationalists' traditional monopoly in matters of culture has been called into question. They have reacted by casting doubts on the authentically founded legitimacy of that 'rival' culture.

The authenticity of identities and traditions appears to be a major concern in both cities in this location. Curiously, this concern feeds into continuing attempts to counteract the retrenchment of a long-running territorial conflict through changing interpretations. When such new interpretations work, their results are considered authentic – and only then. An interesting theme for ethnological research in this regard is the Ulster-Scots culture and identity: is this merely a case of financially motivated opportunism, as its critics claim, or is it an 'authentic tradition'? The answer will depend to a large extent on whether the movement is able to legitimise itself through convincing interpretation, and demonstrate a capacity for development (Kockel 2007b). Everything else – historical facts included – is, in practice, of rather minor relevance.

Voices from the past

This journey in the frontier that is Northern Ireland began at Drumcree, the focal point of one of the most acrimonious disputes about parading in the 1990s. After more than ten years of a 'peace process', parading remains a salient issue in the region. On 27 October 2009, the House of Commons considered a motion by the Democratic Unionist politician Peter Robinson, first minister of Northern Ireland, and 'resolved', following a three-hour debate (Hansard 27 October 2009, Columns 214–56):

> That this House recognises that the right of free assembly and peaceful procession is an intrinsic human right and an important part of the British heritage; acknowledges the cultural significance of parading in Northern Ireland and its tourist potential; regrets the attempts by a

minority to interfere with the right to parade peacefully; and accepts that it is a political imperative to resolve such matters, especially in a context where it is proposed to devolve policing and justice powers to Northern Ireland.

As the minister of state for Northern Ireland, Paul Goggins, reminded the House early on in the debate (Hansard Column 220), '[t]he vast majority of parades in Northern Ireland pass without incident; only a small fraction of the 3000 parades each year are considered to be controversial'. Peter Robinson, speaking to his motion, had quoted more precise figures (Column 218); accordingly, in 2007/8 the parades commission considered 3849 parades, of which 250 – or 6.5 per cent – were regarded as 'contentious', and only 147 of these – or 3.8 per cent of the total – required conditions to be imposed. Laurence Robertson, Conservative member of parliament for Tewkesbury, commented (Column 222) that '147 parades being seen as contentious or difficult and having to have conditions attached to them is a large number, even if a small percentage'. Remarkably, according to Robinson (Column 218), '[o]ne fifth of those "contentious" parades related to the Drumcree stand-off alone'.

The 'parades issue' remains a stumbling block on the road to a social consensus in Northern Ireland. Standing at Drumcree parish church and reflecting on belonging in and to 'Europe', I suggested that until Irish nationalists in Portadown join the Orange procession parading along Garvaghy Road – and are genuinely welcomed to do so – we have a problem. During the House of Commons debate, Iris Robinson, Democratic Unionist member of parliament for Strangford, expressed a desire for 'a day when Catholics and Protestants can celebrate the great pageant of 12 July together' (Column 241), and reminisced: 'That was the case when I was a young girl growing up in Belfast, and I hope that those days will return.' This could be interpreted as a positive signal for reconciliation. Her memories appear consonant with those of Presbyterians especially in the eastern parts, Northern Ireland's Ulster-Scots heartlands, who recall participating in Irish dancing events in their youth. But would that 'great pageant' retain its meaning and purpose? Some reinventions of tradition have been attempted with moderate success, including the Apprentice Boys of Derry's 'Maiden City Festival' around the annual celebrations commemorating the end of the infamous siege in 1689, or more recently the 'Orangefest', an international gathering and parade of Orange Order members in Belfast, commemorating the Battle of the Boyne in 1690. The latter has been hailed as a great success by the city's tourism managers; whether it will, in time, also become a cross-community success remains to be seen.

The elected representatives of the largest nationalist party, Sinn Féin, are not taking up their seats in the House of Commons because this would involve swearing an oath of allegiance to the Queen. Consequently, they could not participate in the above debate but expressed their criticism afterwards. While the issue of territorialism was largely played down by contributors to the debate on the day, the subsequent reactions on the Internet suggest that this issue remains clearly at the heart of the controversy. The Commons motion has become the first move in a contest over the devolution of justice and policing to Northern Ireland. Presenting the motion, Peter Robinson offered a persuasive argument for 'free assembly' and 'peaceful procession' (i.e. parading) as fundamental human rights. He cited (Hansard Column 216) a ruling by the European Court of Human Rights to support the idea that, in a democratic society, those who do not wish parades in 'their' area may say so peacefully, but have no option otherwise but to put up with them. I must leave the finer points of this argument to the legal specialists. It seems that when Sinn Féin says 'we can make progress on parading once policing and justice are sorted out', they are suggesting that, once law and order issues are controlled locally, an equitable solution to the parading issue can be negotiated. When the Democratic unionists say that 'we can make progress on policing and justice once the marching issue is sorted', they are suggesting that once the Orange Order and other groups have permission to march wherever they want, the locally controlled forces of law and order can be put in charge of protecting this as a human right against any opposition. Politically, the first round in the contest had gone to the Democratic unionists who have had their motion – although by no means the underlying issue – 'resolved'. Since then, the devolution of policing and justice has made progress, following a prolonged period of crisis. From an ethnological perspective, it may be observed that the ethnic (Irish) nationalists are using civic arguments that appear inclusive and conciliatory while the civic (British) nationalists – rhetorical protestation to the contrary notwithstanding – are effectively taking an ethnic position that comes across as exclusive and divisive. And so the balancing act continues, along and across multiple borderlines in Northern Ireland.

3
Second Journey – In the Diaspora: *Among Mobile Europeans*

Northern Ireland is one of several ethnic frontiers in Europe where past migrations have created contemporary culture conflict. In the context of European integration, migration among EU member states tends to be no longer regarded as emigration but merely as 'mobility' within the common cultural space of 'Europe'. Implicit in this view is the idea that differences between nation states in this 'Europe' merely constitute regional variants of the same ('European') culture. In the first part of this chapter, I want to reflect on this liberal interpretation of intra-European migration, drawing mainly on ethnographic fieldwork among European migrants in urban centres of Great Britain and Ireland.

The second part of this chapter considers the idea of a polycultural – in contrast to a multicultural – society. The main focus will be on the UK, with some reference to Germany and France. Much of the current debate on migration and culture contact revolves, perhaps understandably, around immigrant groups whose culture is, or is perceived as being, markedly different from that of the host society. But problems of culture contact and conflict are not confined to these groups. Internal migration and the immigration of people from apparently quite similar cultures can prove equally problematic. Migration studies in Europe from the 1970s onwards, strongly influenced by North American theories, have tended to emphasise adjustment and integration. However, this perspective has been recognised as problematic in the context of increasing ethnic mobilisation and a growing emphasis on community and identity. There are many 'invisible' minorities in Europe. Irish migrants in Britain and other English-speaking countries have been described as an 'invisible' or 'hidden' minority for more than two decades (e.g. Grimes 1988). Through multi-sited fieldwork, having lived and worked among Irish migrants in Britain and Germany over some two decades, I had the opportunity to study the transformation of identities and the meaning of 'community' in relation to culture and place among such 'invisible' migrants (Kockel 1999a).

Irish migration to Britain has long ceased to be the extension of an Irish nation as which it is widely perceived, and has instead developed a distinctive culture of its own. The identities of Irish people in Britain are place bound, in terms of either residence (e.g. Liverpool Irish) or origin (e.g. Limerick). Unlike earlier Irish emigrants to Britain and the US, the 'new Europeans' in cities like Munich are perceived as temporary migrants whose mobility is less damaging to, and indeed enhancing, the sociocultural fabric of the communities they come from. Contrary to the stereotype, this mobility is now redefining the concept of 'community' to a point where, analytically, it describes little more than a disparate group of people who happen to be engaged in similar spatial behaviour.

The situation of the 'new' Irish emigrants in Continental European countries is rather different from that of their counterparts in Britain, partly because their numbers remain small by comparison. The social and cultural institutions frequented by these migrants have been created mostly by the host society, or by a major, well-known brewery. Where larger concentrations of migrants have evolved, as in Munich, traditions like Gaelic sports, the Irish language and Irish music in particular are cultivated, often supported by Germans without any Irish connections whatsoever. Migrants and their German friends in Munich used to joke about the day in the foreseeable future when a Bavarian team would contest the All-Ireland Championships in hurling, the Irish national sport.

In the 1980s, the Irish industrial development authority, IDA Ireland, ran an advertisement for overseas investors showing a group of young university students at Trinity College, Dublin, with the caption 'We're the young Europeans'. To the extent that these new migrants are highly mobile, as required by the Common Market, they are indeed young Europeans, and this identity may well be less problematic than that of former Irish migrants in Britain, partly because it remains so vague and ill defined. Familiar expressions of 'Irishness' in Britain, conditioned to some extent by political circumstances, have been backward looking, motivated by the (perceived) need to differentiate oneself from the 'other'. The 'Irishness' of the migrants in Germany is more a 'forward ethnicity', and as such more self-confident and emancipated.

Persistent emigration from Ireland is often explained in terms of 'cultural peripherality', that is, the more or less ready submission to unquestioned exogenous imperatives – a pattern of behaviour said to be the result of colonisation. A migration culture within a wider, largely settled, industrial society is in some sense socially deviant, however economically 'useful' it might be. Herein lies only one of a number of cultural contradictions that an analysis in terms of colonial relationships, with its simple 'stimulus-response' framework, fails to resolve.

Changing migration patterns have been analysed for many peripheral regions of Europe, and comparing the experience of these to that of Ireland could yield valuable insights. The Finnish case in particular would seem to offer a good basis for such a comparison. Both countries were 'underdeveloped' at the turn of the nineteenth to the twentieth century and have suffered large-scale emigration over the past two centuries. Given such an obvious case for comparison, it is rather surprising that little work of this nature has been done to date. Another small peripheral country with a significant history of emigration is Lithuania. Most research on Lithuanian migration has focused on transatlantic migrants (see, e.g. Čiubrinskas 2004); however, when the country joined the EU in 2004, only three member states – Sweden, the UK and the Republic of Ireland – allowed unrestricted immigration from Lithuania, giving rise to substantial migration to these three member states, and this has attracted a number of studies, one of which is briefly referred to later in this chapter.

German migrants in the British Isles[1]

In the light of increasing globalisation, the extent to which a culture is validated can be defined in terms of the reach of its communicative networks (Moosmüller 2000). Accordingly, one's belonging to a particular culture is a function of the frequency and intensity of communication: if you communicate more with group A than with group B or C, you belong to group A. Since ethnic identities appear to become ever more detached from territorial connections of a more traditional kind, it makes sense as a general rule to define cultural belonging in these terms, even though many individual biographies may well highlight the problem of context arising with cultural location according to communication networks. For example, according to this definition I would have been an Irishman during the decade 1978–88 when my social field was made up primarily of Irish migrants in Hamburg, Bremen and Leeds before I went to live in Galway for three years, during 1988–92 in Liverpool I would have been English, then Irish again during 1992–9, German during my time in Bristol 2000–5, and since my move to Ulster

I could be Irish or British, depending on the situation. It might be tempting to view this experience as confirming the popular theory of a postmodern identity warehouse, but I am not convinced. The implicit premises of that theory betray what I would call a shallow essentialism. If an individual can put on or take off a coherent and credible identikit in much the same way as they put on or take off a suit, then the puta-tive essence of the respective identity, which the identikit is supposed to convey, is not disposed of as the theorists would have us believe but merely transferred to the identikit – which may then be fetishised for its metaphysical qualities.

Outside of Germany, German minorities in Europe have been rather neglected in cultural research. Stefan Wolff's (2000) survey concentrates primarily on those groups that previously would have been described as 'ethnic Germans' and on those areas designated geographically and eth-nologically as 'German linguistic territory' and its *Sprachinseln* (linguistic islands) in Eastern Europe. For Great Britain, Panikos Panayi (1996) offered a first overview of the German minority. On the British Isles as a whole, there is a scattering of mostly small local concentrations of migrants with a German background. Some of these local concentrations can look back on a long history as a 'German community' or 'German congrega-tion', even if, in most cases, that history yet remains to be written. From the early 1970s onwards, following the accession of both the UK and the Republic of Ireland to the European Communities, there was an influx of 'drop-outs' and part-time migrants of various descrip-tion, many of whom settled on the 'Celtic Fringe' of these islands, in Ireland, Scotland and Wales. The Germans in the British Isles belong to the 'hidden minorities' referred to earlier because, trying to integrate as best they can, they are as a rule hardly noticeable as migrants. Most of them live in fairly comfortable economic conditions.

In this section I am considering narratives and interpretations of German migrants in contemporary Britain and Ireland, trying to under-stand what, if anything, German culture in these islands involves. This requires a dual reflexivity since, as a participant observer, I can-not simply remain the outsider undertaking research, who intervenes in the everyday life of the migrants 'as if' he were one of them; being perhaps even a typical case of the 'accidental migrant', I am inevitably a subject of research. My research on German emigration to the British Isles emerged originally as a by-product of my doctoral dissertation on regional development and everyday culture in the west of Ireland, and in what follows I make some reference to this earlier work. Over the years, as I turned from student and temporary migrant to become a longer-term career migrant, I began to explore this topic in greater depth, which included an element of self-reflexivity. The discussion in this part of the book draws primarily on participant observation and ethnographic conversations among the German Protestant congregations in Britain and Ireland over more than two decades with some reference to my research among the countercultural immigrants in Galway and other areas of the western seaboard in the 1980s.

In the relationships between migrants and host society it is evident that cultural connections with the homeland continue to exist even where migrants have consciously turned their backs on that country, while cultural connections with their new country of residence have their limits even where considerable efforts are invested to achieve integration. European integration and the globalisation of trade have altered the lived experience of today's migrants significantly in comparison with previous generations. The conscious 'rooting' in the new context nevertheless remains rather difficult. For all the assumed cultural proximity within Europe, it can be shown that within the German cultural experience in the British Isles, spaces and places of concrete everyday belonging are created and find expression most clearly where elements of 'German' culture reach a wider audience.

Historical overview

The history of the German-speaking congregations in the British Isles goes back to early modern times. In London and Dublin, German-speaking Lutheran congregations were already established in the seventeenth century. The 'Hamburg Lutheran Church' in London received its official foundation charter in 1669, at a time, the Restoration period, when other foreign churches in England were rigorously prohibited (Steinmetz 1996). Two pastors ordained at that church founded the first Lutheran congregation in Dublin in 1697 (McCurdy and Murphy 1997). By the time St George's, the fifth German church in London, was consecrated in 1762 there were more than 16,000 Germans living in the British capital.

Religion remained the most prominent external marker of German ethnicity in the British Isles until 1945, in particular during the second half of the nineteenth century, when German migration to England increased significantly and most of the German-speaking congregations were established, many of which still exist today. This conjunction of religion and ethnicity was in part a historical consequence of earlier political practice; during the eighteenth century, the United Kingdom of Great Britain and Ireland, as it was then, had offered refuge to victims of religious persecution from other European countries. This immigration left its cultural imprint on entire districts, for example in County Limerick in the Republic of Ireland, where the linguistic and religious traces of eighteenth century Palatine migrants (see O'Connor 1989; Rasche 1995) can still be detected. In the 1980s, descendants of these migrants established a historical-genealogical society with its own museum to promote this cultural heritage.

A study of religious life in London around 1905 identified some 15 'German' churches but did not include Catholic churches, such as St Bonifaz, in its count. At the same time Manchester had three German churches, and active congregations existed wherever a sufficiently large German population could be found (Steinmetz 1996). The local churches took on far-reaching social and educational responsibilities. In addition, there were numerous associations and institutions which, at least until the outbreak of the First World War, facilitated a rich ethnocultural life.

The war, however, brought about a radical change. Over the years 1914–18, the number of Germans in Great Britain reduced by more than half, from almost 60,000 to less than 23,000. By 1920 there were only nine German churches left in England. Among the German institutions that survived the war were a welfare association, a hospital, an orphanage and, especially in London's West End, a string of hotels and restaurants.

In the 1920s the number of German migrants rose again to about 30,000. As before the war, approximately half of these were living in or near London. There were two German-language newspapers and several cinemas that regularly showed movies in German. The Anglo-German Association was founded in London in 1929. During the 1930s the German population grew further, with estimates varying from 50,000 to 70,000. The census data of 1931 and 1951 do not adequately capture the influx of refugees from the German *Reich* and so exact figures are not available. However, for this period the cultural influence of Jewish immigrants from Germany and Austria is notable, especially in their significant contribution to the development of the arts and sciences in England (Ritchie 1996). These exiles, primarily members of the educated middle class, were often proud of their German cultural heritage. Not unlike the intellectuals who had left Germany after the political unrests in the middle of the nineteenth century, they established 'Little Germanies', for example at Golders Green in the northwest of London. Thus they differed from the majority of those migrants who arrived after 1945 and were more inclined to make themselves invisible, limiting any expressions of national pride to the quality of goods 'made in Germany'. Many of the German institutions that survived the Second World War go back to Jewish foundations laid during the interwar period.

In the early years after the Second World War the German population in Great Britain grew once again considerably (Kettenacker 1996). There were some 15,000 men, prisoners of war who stayed on in the UK, although not all of them permanently. Then there were a large number of women, made up in the main of three groups. Some came as au pairs. Once the universities in Germany were operating again, there were a growing number of student teachers keen to experience a foreign culture. There were women who had married British soldiers and returned with their husbands – *Beutebräute* (captured brides), as many of them self-mockingly called themselves (see Bindemann 2004). A further group of migrants from Continental Europe who arrived in Great Britain after 1945 were the expellees, some 90,000 of whom were employed to help with rebuilding the British economy; approximately ten per cent of these migrants were *Volksdeutsche* (ethnic Germans) from the former eastern territories of the *Reich*. A relatively large number of migrants in this period were 'stranded' in Great Britain – individuals who had not intended to migrate but had been made emigrants by circumstances. To represent and protect the social security interests of all German immigrants, the *Deutsche Wohlfahrtsausschuß* (German welfare committee) was established in 1952.

A major proportion of the German population in the UK nowadays consists of women who have acquired British citizenship through marriage (see Steinert and Weber-Newth 2000). They are scattered widely throughout the realm, with the majority living outside of London. Many have overcome the social isolation of their early years in these islands through intensive participation especially in the German-speaking congregations. Their British husbands have often joined the congregation and have also become actively involved. All census returns for German-born migrants in the second half of the twentieth century show a notably larger number of women than men, although the discrepancy is less clearly marked now than it was in 1961, when the ratio was approximately two to one. The total number of inhabitants of Great Britain who were born in Germany is shown in the 1991 census as 215,113; by 2001, this had increased by almost 22 per cent to 262,276. A considerable number of these would be children born into families of British forces personnel who returned home after German unification had brought the end of the post-war military occupation of Germany. Indeed, four of the five largest concentrations – Wiltshire, Colchester, North Yorkshire and Aldershot – are major army bases and show a combined German-born population of some 12,000. Even with this factor taken into account, however, the Germans remain the second largest 'white' immigrant group after the Irish, and ahead of many nationalities from the Commonwealth countries.

From about the mid-1950s, prompted by a progressive Anglo-Americanisation of the global economy, the incidence of short-term career migration increased sharply. Since the 1980s, following the introduction of intra-European educational mobility programmes, the number of transient migrants also includes a growing number of third-level students coming for up to a year, and sometimes staying on. For these transient groups, London is clearly the key destination, although other urban centres also fare well, and the dominance of the British capital is less pronounced nowadays than a century ago: in 1991, some 34 per cent of German-born migrants lived in London and southeast England, with less than half of these in the capital itself, and the percentage remained stable in 2001, with a slight intra-regional shift towards London. Outside of the Greater London area, the German congregations have remained throughout the second half of the twentieth century as the most important platform for the reproduction of German identities in Britain, in spite of financial difficulties leading to a reduction in the number of *Pfarramtsbereiche* (parishes covered by a pastor; nowadays usually including several congregations).

In Ireland the situation is somewhat different. Around 1800, there were a number of rural areas where German – or, rather: particular German dialects, such as Palatine in County Limerick or Hannoveran around Athlone in the Irish midlands – was the vernacular language of entire village communities, albeit for relatively short periods of time only. Although there was some German immigration to the island of Ireland throughout the nineteenth and early twentieth centuries, this lacked the numerical significance of French or Italian immigration, which attracted slightly more interest among researchers. Outside of Belfast and Dublin, the congregations are relatively small. The pastor is based in Dublin, and since 1984 the entire island forms a single *Pfarramtsbereich*. At the end of the 1980s, I had been one of the first researchers to describe contemporary Ireland as a country of immigration (Kockel 1989a, 1991), but Irish migration research, for long fixated on mass emigration, only began taking this topic seriously after attacks on asylum seekers and other foreigners coming to live in Ireland could no longer be ignored.

Observations on location confirm that the countercultural migrants and other 'drop-outs', who from the late 1960s onwards settled especially on the western seaboard, still constitute relatively large groups within their respective local area. At the time of my field research there during the 1980s there were several villages along the coast from County Donegal to County Kerry where, even outside the short tourist season, one could hear almost as much German as English. That immigration was decisively influenced by the broader 'Folk Revival' in Continental Europe; with the decline of that movement since the 1990s, the social and cultural profile of German migration to Ireland is likely to have changed.

As already indicated, my research on German immigration to the British Isles emerged as a kind of by-product of a dissertation on regional development and culture in Ireland, undertaken at the University of Liverpool. I had already been aware of the existence of German congregations since my undergraduate days in Leeds, where Lutheran House, a religious and cultural centre for immigrants from the Baltic, Nordic and German-speaking countries, was located just across the road from where I lived. At Liverpool, the German church is located only one street away from the university precinct. Initially and for some years a participant only, I became a participant observer when, after the 1991 census, a press report (which I have unfortunately failed to locate again since) alerted me to the possible status of German migrants as one of the largest – yet hardly researched – migrant groups in Britain.

While the German congregations provide the main ethnographic setting for the following explorations, these explorations are not limited to them. I am not attempting to offer a representative sample of German culture in these islands; instead, I am reflecting ethnographically on a particular cultural context, with the emphasis of description and discussion on the experiences of contemporary migrants, a spectrum that covers several generations and reaches back to the interwar years.

The German-speaking congregations

In the mid-1980s, when I discovered the little church near the university on Liverpool, there were still more than a dozen German pastorates in Britain, organised in their own synod. Over the following decade or so, the financial constraints of the synod, and of the Protestant Church in Germany (EKD), which used to support it until recently, made the consolidation of groups of long-established congregations into seven *Pfarramtsbereiche* inevitable. Some congregations with a long tradition decided to leave the synod as a result, while other congregations suffered an exodus of younger members. In the following paragraphs I introduce those congretations on which my field research has primarily concentrated.

Liverpool (North and Northwest England/North Wales)

The congregation in Liverpool is one of the longest-established German communities in Britain. Today it is one of the few congregations that still have their own church building, even though the pastorate has long been based in Manchester, which is the administrative centre of the new *Pfarramtsbereich Nord-und Nordwestengland/ Nordwales*, which stretches from the North Sea coast in Yorkshire to the Irish Sea at Anglesey. In irregular intervals, student pastors may come to spend some time working in the *Pfarramtsbereich* as vicars, and before the 'eastern enlargement' of the *Pfarramtsbereich* they would have usually lived in Liverpool so that the congregation would 'as good as' have its own pastor every so often. I was a participant observer in this congregation between 1986 and 2001; given the length of time and the depth of knowledge and understanding I gained there, it forms the benchmark case study from which primary insights are drawn and which generates key questions to explore in the other congregations. The congregation has an active cultural life beyond its religious dimension, and belongs to the livelier congregations in which all generations are reasonably well represented, even if the majority of active members, here as elsewhere, are from the older age groups.

Bristol (Southwest England/South Wales)

Bristol was one of the last old and well-established congregations that lost its own pastor. The pastorate is now based at the Welsh capital, Cardiff, an hour by car on a good day, where the congregation has the use of a Methodist church for its services. In the south of the *Pfarramtsbereich*, the Bournemouth congregation still has its own church building and a 'subsidiary' pastorate and up to the early 2000s was staffed regularly by a vicar. In terms of demographic characteristics, the Bristol congregation appears relatively old compared to its counterpart in Liverpool. Services take place in the parish hall of an Anglican church and were held twice a month during my period of field research, with the pastor and a church elder who was trained in ministry taking turns, but have been restricted to once a month since the elder retired and moved to France. There are also occasional family services and special celebrations that are held at other locations in the Bristol area. Social gatherings take place once a month at the same Anglican parish hall, and a Saturday School and playgroup has been active for some time, which signals a rejuvenation of the congregation. My participant observations here extended from 2000 until 2005, followed by regular mail contact.

Edinburgh (Scotland/Northeast England)

Edinburgh is the largest of the Scottish congregations and one of the oldest in these islands. Pastorate and church building are located in the same grounds, and as in Manchester and now in Bristol, there is a German Saturday School. The congregation is relatively large and has the most balanced age profile of the congregations described here. My observations in Edinburgh have been sporadic since 2003. The pastor at the time, Dr Walther Bindemann, had come to the congregation in Newcastle in 1995, when Northeast England was still a separate *Pfarramtsbereich*, and had moved to Edinburgh when it was amalgamated with Scotland. His collections of bio- and ethnographic interviews with members of the various congregations that make up the new *Pfarramtsbereich* offer valuable insights into these migrant communities (Bindemann 2001, 2004).

Belfast (Ireland)

The German congregation in Belfast was originally affiliated to the German synod in Britain until 1984, when it officially became part of the Lutheran Church in Ireland. It is the 'little sister' of the congregation in Dublin, and the only one in the island of Ireland outside Dublin to enjoy a regular monthly service. Whereas the congregation in Dublin along with the pastorate, a school and other facilities also has its own church building and community hall, the congregation in Belfast meets at a Moravian church. By reciprocal arrangements, the Moravians in Dublin are using the German Lutheran church. The age profile of the congregation in Belfast is similar to that in Liverpool, but the congregation appears much smaller. This may be due to the differences in regional population geography. Both congregations are situated in a regional population of about 1.7 million, but Liverpool/ Merseyside as a metropolitan area has a much greater population density and better public transport infrastructure, which makes it easier for members who live some distance away to attend services and other activities regularly. My participant observation in Belfast began in the spring of 2005.

Outside the congregations

The experiences of members of the congregations can be compared with those of other immigrants who locate themselves culturally without any reference to these congregations. Among these other migrants, I have frequently met individuals who have consciously turned their back on Germany and have been making every effort to integrate in the host society. Observations and interviews relating to their experience that are informing the following discussion have been conducted primarily in Bristol, Liverpool and London as well as in both parts of the island of Ireland.

Observations and other insights

A handful of themes may be identified that extend across the different generations of migrants. These include in particular issues of language and communication in the widest sense, as a process formed by values and patterns of behaviours that have their roots in the childhood of the individual. This applies not only with regard to feast days and holy days in the annual as well as the individual life cycle but equally in everyday life: from table manners to ways of greeting, leisure habits or ideas and rituals of cleanliness. Habitual attitudes and patterns of behaviour become problematic when they lose their casualness in confrontation with another, foreign lifeworld.

Although officially German speaking and Protestant, the congregations discussed here include among their members others, such as French-speaking Huguenots from Switzerland, Catholics from Limburg in the Netherlands and even Old Catholics from Bavaria. What they all have in common is that, despite their religious and linguistic differences, they wish to maintain a spiritual connection with Continental Europe through this link with 'German' culture. Such cultural affinities can be observed even where migrants are consciously rejecting their country of origin. In this regard, three groups can be distinguished: 'drop-outs' in search of an alternative lifestyle; the immediate post-war generation, many of whom reject any reference to national or ethnic specificity; and the mobile inhabitants of the global village with its identity warehouse. What these three groups have in common, however, is that despite their ostensibly negative attitude towards their country of origin they maintain cultural connections with it. This is particularly apparent in three spheres of everyday life: in their media consumption, in their search for contacts with other Germans, and in eating and drinking.

Until fairly recently, it was difficult to find German newspapers and magazines outside the tourist season except in the biggest cities. In Britain migrants frequently cite the anti-German attitude of especially the tabloid press as the reason why they prefer German media. Before the spread of satellite television in the 1990s, many would also listen to German radio on the AM wave band. Not long after television arrived via the Astra 1 satellite, migrants noted with bemusement that virtually all British channels moved to a separate satellite location, so that it was no longer possible to receive both British and Continental European channels with the same basic dish and decoder. This was interpreted as yet another incidence of British 'splendid isolationism'.

Continuing cultural affinities are also visible in the endeavours of many migrants to make and maintain contacts with other Germans. In the west of Ireland, for example, the vigorously displayed aversion to German holiday makers – with a derogatory German neologism referred to as *die Touries* – goes hand in hand with a range of contradictory behaviour, from strategically 'hanging out' at particular popular tourist spots – with the aim of speaking a few words of German or to catch the latest news from home – to involvement in a cottage craft industry that lives mainly from tourism, which in many locations on the west coast comes predominantly from Germany. Networks are an important factor of identity creation, especially when their activities are concentrated on events of symbolic significance. That includes the annual cycle of religious and other feast days, especially during Advent, and also other regularly recurring events that allow the migrants, quite regardless of their declared aversion to their country of origin, to come into contact with 'German' culture.

A third area, perhaps the most important one, is eating and drinking. German migrants are by no means unique in this regard. Student projects I supervised in Bristol (2000–5) regularly found that immigrants from the Mediterranean region received food parcels from relatives at home, even though Mediterranean specialities are nowadays widely available not only in supermarkets but also in an array of ethnic and regional specialist delicatessen. German palates outside London are less well served, but some German food is now appearing in the major supermarkets, and consumers are no longer fed the scaremongering myths of the 1980s when at least one of these supermarkets advised its customers that Cervelat sausages or Black Forest ham are raw meat and must only be enjoyed after having been cooked for a long time and at fiercely high temperatures. Since the Common Market made inroads into the grocery market in these islands in the 1990s, discount supermarkets like Lidl or Aldi have further helped to reduce German withdrawal symptoms, and some of the delicatessen based in the London area now offer mail order services via the Internet. After 2004, with the eastern enlargement of the EU, numerous Polish and Lithuanian food shops and even some restaurants opened throughout these islands, and many of these stock goods that are, or are 'as good as', German (including traditional Russian and Georgian cheeses manufactured by *Rußlanddeutsche*, ethnic Germans from Russia, in a small town near the Black Forest). Before this culinary revolution, even migrants with limited financial means would regularly undertake pilgrimages to the few shrines of 'German' food culture, ranging from a butcher in Killarney in southwest Ireland to an eastern European bakery that had a franchise in Lewis's department store in Glasgow. It is interesting to note in this context that already in the late nineteenth century food was a key marker of German culture in these islands. Rosenkranz (1965) points, for example, to the prominent role of German immigrants in the confectioners' and pork butchers' guilds in Liverpool. For the US, Louise Erdrich (2003) has given this image of the German migrant literary expression in *The Master Butchers Singing Club*. Apart from the variety of sausage types and some regional specialities, such as plum jam ('made from plums rather than sugar') or potato dumplings, what German migrants in these islands miss more than anything else is *anständiges Brot* – 'decent bread'. That goes for the postmodern part-time migrants every bit as much as for the more settled immigrants. When I ask students to bring along to the seminar something that represents their culture and identity, British or Irish students may bring their mobile phone or their credit card – however incredible that may seem – while German exchange students time and again will bring bread, 'because you can't buy decent bread over here'. In the *Pfarramtsbereich* based at Manchester, the pastor regularly used to bring to the various congregations baskets full of bread from a bakery near the pastorage that specialises in serving Polish and other Central European palates, and in a London suburb a German bakery recently evolved into a mail order delicatessen. This company also offers a range of German flours milled to different grades that are not available in ordinary shops or supermarkets in these islands. Trying to bake German-style bread with British or Irish flours invariably leads to frustration as these flours give a different texture.

Mobility is a key feature of our time. Within Europe at least, it is relatively easy today to change your place of residence. Moving to another European country in search of work is no longer unusual. Budget airlines and communication technologies open up new possibilities for staying in touch with relatives and friends in the country of origin. And yet inevitable intercultural tensions arise and determine the framework within which migrants need to negotiate their personal identity. That identity is, on the one hand, subject to more or less subtle changes; on the other hand it can be very important for individual migrants, or indeed for migrant groups, to project, emphasise and maintain a stable identity element. Next to the biological and genealogical aspects acquired through birth, these building blocks of identity include primarily aspects we acquire in the course of our life – linguistic ability, knowledge, skills, understanding – and which become the foundations of our image of ourselves and the world around us. An important constitutive element of all identity is therefore also the 'other' that we encounter (Kapuściński 2008). Migrants' experience of the 'other' differs in obvious ways from the experience of non-migrants. In the interaction of 'own' and 'other', the personal identity of the individual is thus modified. Moving to another country can bring particularly stark changes.

Identity is invariably detemined by the experience of boundaries and frontiers. German migrants in the British Isles move in certain frontiers and along and across certain boundaries – they speak two (or more) languages, have life habits, customs and traditions that distinguish them from others in the same local area, which often includes members of their own family.

Since the 1990s, satellite television and the expansion of the international traffic infrastructure have made it much easier for emigrants to stay in contact with their country of origin. The food situation has improved, thanks to the internationalisation of trade – although the conversation between two Germans meeting for the first time in these islands still only takes a few minutes before it turns to the inexhaustible topic of 'decent bread'. It has become much less complicated than only a few years ago to identify oneself culturally as German. Moreover, it has become easier to feel 'Irish', 'English', 'Scottish', 'Welsh' or indeed 'British', or to alternate freely between a globalised version of any of these and an equally globalised German identity. And yet the postmodern identity sunshine, forecast to bring about the dissolution of identities in some multicultural 'melting pot', has not materialised.

The *Heimat*-question

When German immigrants talk about their identity, they often use the term *Heimat*. Many migrants have lived in these islands for a long time, often longer than they ever lived in Germany, and have children or grand children here. On the one hand, there is a kind of indeterminate longing for another *Heimat*, which resonates in many life stories and interviews (see Bindemann 2001, 2004). It also becomes clear through observations and conversations that *Heimat* is not something given and predetermined but something that is only appropriated through hard and sometimes painful identity graft. Moreover, one can be 'at home' in many places simultaneously. *Heimat*, however, as it resonates in many of the life stories of migrants, is something else. Among the younger Germans who have come to live in these islands, the concept of *Heimat* has increasingly become a topic since the late 1990s. Remarkably, the meaning of the term for 25-year old migrants differs little from its meaning for the 75-year olds, and for both groups it contains something that 'shines into the childhood' (cf. Bloch 1978). For a woman belonging to the younger generation, *Heimat* is no longer an old-fashioned term associated, as it was before she emigrated, with schmaltzy *Heimat* films, 'mountainous forests and roaring stags'. It denotes, on the one hand, the rediscovery of everyday cultural patterns from her own childhood: 'Just as my parents did, my husband and I nowadays shop at Aldi; we have a black-and-white telly and no car.' Her husband is English, and thus *Heimat* is, on the other hand, the fusion of two cultures within a family, 'where we keep common as well as different traditions, and understand them in new ways'. Members of the older generation emphasise the role of childhood and youth for *Beheimatung*, the location of the individual in relation to *Heimat*. They also stress that *Heimat* is not so much an association with any specific place but rather particular activities. For a woman who was born in Berlin between the wars, this meant less the city itself, but *das ganze Drumherum* – everything related to it that had personal significance. In her case, that included the Spreewald forest south-east of the city, where she spent much time on a farm owned by friends of the family, 'picking mushrooms and blueberries, or going for a swim in the clear water of the lake'. Language, too, forms part of it – not only the standard 'High German' but especially accents and dialects. When the English husband of a German woman who had been expelled from Pomerania as a child introduced her to the local congregation in the 1950s, she felt, 'quite spontaneously and without rational explanation', a strong sense of belonging in spite of her rather critical attitude towards the institution of the church. There was 'the language of *Heimat*, which I heard for the first time in many years, and which I had almost given up, in favour of English'. Not only the spoken language, non-verbal communication can also be important, as a middle-aged woman put it: 'those other signals, the ones you only know if you have grown up somewhere'. Even after several decades in the new environment, their lack of understanding of such signals can 'show up' migrants as 'others' and the resulting feeling of being a stranger can be hard to overcome.

For many the notion of *Heimat* is connected with memories of childhood and youth. When they return to the world of their childhood today, they often experience that *Heimat* as something 'other', a foreign place. This is especially, understandably, the case with migrants from the former eastern territories of Germany. Individuals who have moved around a lot tend to be able to find a kind of 'mobile *Heimat*', in the sense of belonging to a particular group, not only in family traditions but also in personally meaningful songs or ideas that they can take with them wherever they may be.

In the course of discussions about *Heimat* it is often claimed that British friends and acquaintances simply cannot understand the problem because they lack the necessary terminology. An older woman, who had moved several times and lived in different countries, said that she could feel 'at home' anywhere where she could live happily and comfortably, but 'that isn't at all what *Heimat* is about'. England, where she had been living for more than 30 years, was no more *Heimat* for her now than Germany. It was for her '"home", as the English would call it', but that is 'an enormously different matter'.

In Germany itself, *Heimat* was debated vigorously during the 1990s, with many authors lamenting the disappearance of *Heimat* (e.g. Hecht 2000) as a consequence of globalisation. Whether the simultaneous debate among the migrants simply reflected the zeitgeist in Germany, or whether this debate was qualitatively different, must be left to further analysis. It should be noted, however, that the discussion in Britain mainly emerged from and was conducted within various initiatives associated with the congregations, and in which the different generations of migrants meet. Initiatives such as religious house groups, or events such as the Advent bazaars, are important well beyond the active membership of the congregations. Complex traditions are first outlined in these contexts – regardless of whether they correspond to any traditions that were kept within an individual's lifeworld prior to emigration from Germany – and can become for individual migrants pillars of their new personal identity. Once again, food plays an important role in such gatherings – sausages with potato salad at the Advent bazaar, new potatoes with quark for the harvest festival or pasta salad after Bible study.

Even in a globalised world, people who come from another country remain 'others'. This includes German migrants in these islands, even if they have been living here for a long time and have become relatively well integrated. It is evident not just in their food habits or certain customs and traditions but especially in communication with the local host society. Communication processes follow group specific, unwritten rules that create a common ground. Where this foundation is not there, the individuals concerned realise how much their ways of thinking, feeling or speaking differ and make them 'others' despite all the superficial trappings of integration. For German immigrants in Britain, an added complication is that they have to come to terms and live with clichés and prejudices that are propagated not only by the tabloid press. In this context, Bindemann (2004) points out that media-driven rabble-rousing can only blossom where it finds fertile ground.

From observation and interviews it became evident that individual migrants are attracted to the congregations, and to the opportunities they create for *Beheimatung*, not so much by religious observance or good sermons but by the social structures and their development potentials. Time and again, the function of the congregations as primarily social networks is emphasised, independent of any matters of faith. In the light of increasing secularisation, the question arises whether these networks will work in the long term. Considered in isolation, the number of people attending church services could give the impression that almost all congregations are facing extinction. When the full spectrum of activities in the congregations is taken into account, a more differentiated picture emerges. In the course of the anniversary celebrations in several of the congregations around the turn of the twenty-first century, it was pointed out more than once (e.g. Bremer 1996) that the German congregations in Britain have always been 'congregations of the first generation' and thus 'on the brink of extinction'. This is a consequence of, among other factors, the large number of 'mixed marriages', where children grow up between two cultures, but – especially where everyday language use is concerned – practically identify more with their British context and as adolescents often lose contact with the respective congregation. Notably many of these children return to the local congregation as adults, sometimes at a different place of residence, once they have children of their own.

It has already been noted that the congregations are not composed exclusively of German Protestants. Even if the general culture contact with Catholics, atheists or adherents of other faiths from Germany is limited – German-speaking Catholics, for example, have their own ethnic infrastructure, centred in London and Dublin – there are extensive connections at the individual and group level. In some locations German reading clubs have been established and the larger ones of these offer a wide range of events. Since 2008 there is once again a small German-language newspaper, *Germanlink*, carrying an extensive list of events and other information of interest to German migrants. It is quite conceivable that, in the course of time, secular initiatives such as these may replace the congregations as primary institutions for the construction and cultivation of German identity.

While during the early periods German ethnicity in the British Isles was primarily determined by male interests – in the congregations as much as in trade associations and other organisations – this changed after 1945. Women have increasingly taken on the initiative, not only in the congregations. Most of these women are or were married to British husbands. There are also local 'activists', some of whom have no immediate family connections with Germany but who may feel an affinity with German culture for other reasons. One of these, a middle-aged Englishman, who supported German cultural life in his area over many years, ended up being elected as an elder of the local congregation.

In contrast to immigrants in the nineteenth century, and also to the mainly Jewish refugees in the 1930s and early 1940s, today's German migrants are not creating any 'little Germanies' in the sense of entire streetscapes, or urban or rural districts, with a distinctly German character. An exception is a residential area in the west of Greater London, popularly referred to as 'Germany plc', where there is a high concentration of embassy staff and employees of German companies and organisations.

As already indicated, it has become much easier over the past two decades to be German – or whatever else – in Britain, Ireland or anywhere else in the western world. The everyday experience of German migrants in these islands is full of what in practice are cultural 'third spaces', including the churches and other rooms used by the congregations. The German butcher in Killarney offers one of these spaces, as does the bakery in Twickenham near London or the café 'Im Backstübl' in Warrington near Manchester. So-called theme pubs with silly names, such as 'Beer Keller', are not part of this and are rarely frequented by German migrants. By contrast, the 'Oust House' in Southport, north of Liverpool, does not advertise itself as a German pub, but its atmosphere makes it immediately recognisable as such once you step inside. That has something to do with a sense of the usual – the German word, *das Gewohnte*, has a Heideggerian resonance of 'having been dwelt (in)' – and with non-verbal communication. Anyone entering such a 'third space' recognises, even before the ear picks out the familiar sounds from the babble of voices, by gestures and mannerisms, dress sense and other signals that those 'Little Germanies' do exist still. Only, they are no longer streetscapes with an unmistakably German imprint, but rather scattered places where people come together. Some of these, such as the 'Backstübl' mentioned earlier, signal their identity clearly, others, such as the mobile take away in York with its gourmet German-style *frikadellen* meatballs, do not.

What all these places have in common is that they are not some folksy stylised piece of Germany in these islands but something different: German places as they can only exist here – no in-between thing, no hybrid. They are at once entirely German – in much the same way as Bavarian or Saxon is at the same time also 'German' – and entirely English/Scottish/Irish/Welsh or, for that matter, Ulster. There is little cultural fixation, whether by people in their everyday lives or by politicians, and not even by that handful of ethnologists and historians who are interested in these migrants. Germans in the British Isles constitute an ethnic minority, but they create their *Heimat* in everyday interactions with one another or with the host society. Arising from these everyday interactions is their specific identity, whose roots in the country of origin are both more and less pronounced than they appear to have been for earlier generations of immigrants. To unravel this apparent contradiction by comparative research with other immigrant groups would be a rewarding task for further field research.

Monocultural policy on the way to a polycultural society[2]

In the previous chapter, I considered the relationship of identity, nationality and citizenship in the context of 'settled' cultural groups with conflicting territorial claims and interpretations. Having discussed migrants in the present chapter and raised the issue of belonging – which I will return to later in the book – I now want to take a closer look at the relationship between citizenship and cultural identity, especially in the context of cultural policy. In public discourse, citizenship is, on the one hand, often implicitly treated as synonymous with cultural identity; on the other hand, it is also understood as superordinate to identity. Staying focused on Britain, I want to explore this ambivalent relationship a bit further.

Since the union of the Crowns of England and Scotland in the seventeenth century, an ideal-type of 'Britishness' has been constructed as *the* multicultural citizens' identity par excellence. Thus while *citizenship* is generically 'British', a person's nationality may be, for example, 'Scottish' and her or his identity – understood as an entirely personal, individual matter – may be, for example, 'Catholic', 'Black' or 'Lesbian', or any combination of such self-ascriptions. In this 'nested hierarchy', the concept of 'citizenship' clearly constructs an overall category as inclusive, and thus an invariably positive self-ascription, while 'identity' characterises diverse minorities as exclusive and thus at least potentially a negative self-ascription. As indicated in the previous chapter, the continuing tensions in Northern Ireland after more than a decade of a 'peace process' reflect the problematic of these constructs in relation to established residents; with regard to immigrants, especially those from Commonwealth countries, the same problematic has been debated within 'British Studies' since the 1960s (see Bassnett 1997; Storry and Childs 1997). Until not so long ago, immigrants from these Commonwealth countries would automatically be British citizens, but nowadays, in spite of the popular policy rhetoric of integration, the same migrants are increasingly feeling alienated (Modood 1992; Commission 2000; Wazir 2002). In recent years, the future of Britain as a multi- or poly-cultural society has been called into question not least through opinions expressed by cabinet ministers who appear to want to relegate non-Anglophone cultural expressions to the private family sphere. Parallels with the historical 'naturalisation' of the 'Celtic Fringe' are critically noted by contemporary immigrants, for example at an 'Ethnic Minorities Forum' that formed part of a research workshop on European ethnology, held in Belfast in June 2002.[3] The fact that such parallels are drawn may suggest a degree of assimilation.

In the following paragraphs, the policy of the Labour government since 1997 will be outlined against the background of some demographic data on immigration to Britain.[4] The problematic raised above appears especially clearly in the debate on the cultural, social and political implications of an increase in the Muslim population, a debate that revolves primarily around the two poles of culture (in this case religion) and politics (in this case terrorism). Later on in this section, I discuss how culture, and language in particular, is politically represented as a channel for mediating citizenship and identity, before considering some paralles and contrasts with the situation in France and Germany, and asking whether – and if so, to what extent – the vision of a monoculturally constituted polycultural society can be realised.

Immigration and integration in Great Britain

In the British census of population for 1991, 'ethnic' data were collected for the first time, and the 2001 census was the first to include 'mixed-race' data. According to these data, almost half of Britain's 'ethnic minority' population are living in London, where some 300 languages are spoken in the schools (Katwala 2001). Central London is the only part of the country where 'Black Britons' outnumber 'British Asians'.[5] 'British Indians' are the dominant minority group in the London suburbs. The suburbs Newham and Brent are the first districts of Greater London with a 'non-white' majority population. Towns and cities in Northwest England that have seen frequent ethnic tensions and open conflicts in recent years display a high degree of ethnic segregation. In Rochdale, for example, 96 per cent of all Pakistani immigrants and 89 per cent of all Bangladeshi immigrants are concentrated in five central districts, which are among the most deprived districts in the entire region. Pakistani migrants form the largest group in Northwest England, in the three counties of Yorkshire and also in Scotland, while they are relatively underrepresented in London. In the Midlands, as in most predominantly 'white' regions of England, immigrants from India are the largest group although it should be noted that the census does not distinguish between the diverse ethnic groups from the Indian subcontinent. Similarly, there is no differentiation between ethnic groups from China; the Chinese immigrants are widely dispersed across Britain and hardly feature in public debates on immigration. Indian immigrants are on average better off financially than the 'white' population, but social differences within the group are more pronounced. Pakistani and Bangladeshi migrants constitute the poorest groups and fare badly with regard to virtually all indicators of social exclusion and disadvantage.

In the medium and longer term, immigration brings about demographic shifts that go beyond the immediately measurable population growth. A comparison of the age medians of various groups indicates some changes that may be expected in the future (Table 3.1). These figures suggest that the share of ethnic minorities in the British population is likely to grow even without further immigration, as the children of migrants grow up and establish families of their own.

The Parekh-Report (Commission 2000), which was published before the 2001 census, deduced from its data sets some demographic projections according to which the expected overall growth in the British population

Table 3.1 Age median of ethnic groups

White	37
Indian	31
Bangladeshi	18

Source: Katwala (2001)

Table 3.2   Ethnic minorities (estimated in thousands)

Group	1998	2020
African	354	700
Afro-Caribbean	797	1,000
Bangladeshi	232	460
Chinese	167	250
Indian	945	1,200
Irish	2,092	3,000
Pakistani	567	1,250
Various	601	1,000
White (except Irish)	50,986	49,000
Total	56,741	57,860

Source: Commission (2000: 375)

will be attributable to immigrant ethnic minorities while the share of the non-Irish 'white' population will decline (Table 3.2). Even though the largest share of the shift suggested by these projections results from the assumed growth of the Irish migrant population, the figures point towards an ethnically increasingly heterogenous Britain.

In the aftermath of their landslide victory in the 1997 general elections, the Labour Party, then under the banner of 'New Labour', initiated a project conceived of as the societal modernisation of the UK. While emphasising a modern national identity, a high regard for ethnic diversity was purposefully highlighted. The public celebration of Great Britian as a 'multicultural society' became a leitmotif of political rhetoric. Among the symptomatic efforts to popularise a new British sense of identity that would incorporate the cheerful assimilation of new cultural goods and practices was Foreign Secretary Robin Cook's often cited description of the 'Asian' dish 'Chicken Tikka Masala' as the British national dish. The ensuing debate challenged the authenticity of the dish, which is widely regarded as a concoction of 'Asian' chefs to please the British palate – an observation that gave scholars of a post-modern disposition copious opportunity for discourse. At the same time, and even more so in the aftermath of the urban riots in Northwest England during 2001, the political position of 'New Labour' in relation to the problematic of immigration and integration appeared to be Janus headed. Prime Minister Tony Blair's assessment of the unrests as a 'law and order issue', the draconian sentences handed down to many young 'Asians' who had no previous convictions and, not least, the Home Secretary David Blunkett's public criticism of the campaign for greater leniency were seen by many as planting the seed of future unrest (Singh 2002). Studies of the riots point to a lack of social cohesion, a situation in which there is little contact between the majority population and the minority. The Cantle-Report on Oldham (a suburb of Manchester), whose conclusions strongly influenced the proposal for new legislation on migration and citizenship, has been described as being full of platitudes on the topic of migrant integration, and thus diverting attention from the actual problems that young people, quite regardless of their ethnic background, are confronted with in that area – the under-resourcing of schools and social infrastructure, poor quality housing or a high level of unemployment, to name but a few (Werbner 2002).

In the new cabinet after the second election victory of 'New Labour' in 2001, the former education secretary, David Blunkett, was made home secretary, and in this capacity he affected a change of political emphasis away from multiculturalism. The primary focus was now to be on the creation of a stronger, more coherent national identity. In the public debate that followed, the fact that such an attempt feeds back onto the tensions between the individual constitutional parts of the UK was not widely recognised at first, as Blunkett's advances in relation to immigration restrictions and societal integration aroused much greater, more immediate interest. Particularly controversial was his idea of a 'Britishness' test, designed to establish the aptitude of migrants to participate in British social and cultural life. In the aftermath of the unrests of 2001, public discourse revolved primarily around the integration of diverse cultural and religious identities within a framework of values shared by all residents.

In the Foreword to *Secure Borders, Safe Haven: Integration and Diversity in Modern Britain*, Blunkett wrote about the need for a strong sense of belonging and identity on the part of the majority population, describing them as the indispensable foundation for integration and healthy diversity. 'New Labour' was increasingly playing on a reviving national pride, the core of which appears curiously archaic. The draft legislation was described in the *Guardian* newspaper of 8 February 2002 as an 'almost modern' approach to the immigration issue, since it recognised immigration to Britain as part of a larger global phenomenon. In that sense, the paper regarded the draft legislation as a welcome turning point, as it seemed to place the system of state controls in the service of the social integration of cultural diversity, rather than trying to uphold an increasingly imaginary, monocultural status quo. However, critics such as the anthropologist Pnina Werbner (2002) have noted that although the state claimed a commitment to diversity and multicultural policies, at the same time it continued to define citizenship implicitly in cultural terms – especially with reference to language and religion. This was one reason why a growing number of critics, such as Burhan Wazir (2002), pointed to the growing alienation of older migrants who settled in Great Britain some time ago but are feeling increasingly unwanted. This mounting sense of alienation among older immigrants has its origins not just in the most recent wave of Islamophobia but indeed goes back to the draft legislation debated in parliament in the autumn of 2002, which had been in process long before 11 September 2001. When he was education secretary, David Blunkett had already introduced citizenship education in schools, and from September 2002 this was made a compulsory subject at secondary schools. It is not without irony that children are taught about many different cultures in the course of their citizenship education – except about English culture (Alibhai-Brown 2000).

Some 80 per cent of British Muslims are from the Indian subcontinet, in particular from Mirpur and Punjab. Kinship connections, so-called brotherhoods, play an important role among these immigrants (Shaw 2002). These brotherhoods support the creation of a parallel society and economy with strong socio-economic and emotional ties to the country of origin, which are relatively disconnected from the host society. Such parallel 'ethnic economies' are by no means limited to the poorest groups among the immigrants. A lack of cultural awareness is cited as a major reason why the take-up of state-funded support initiatives is low among immigrants (Gidoomal et al. 2001). To this day the Mirpuri and Punjabi migrants from Pakistan, together with migrants from Bangladesh, belong among the social groups featuring the highest proportion of households with below-average incomes and the highest rates of unemployment. Thus their situation contrasts sharply with that of later immigrants from East Africa; these include a large number of Gujarati Muslims of Indian origin, many of whom have come with savings and qualifications. The reasons for this disparity lie partly in structural exclusion and partly in cultural factors. Men from those Pakistani groups have tended to remain in unskilled occupations much longer than their Punjabi Sikh or Hindu counterparts, and have persisted in the practice of 'international commuting', which endangered their jobs because of their extended visits to the homeland (Shaw 2002). That strong link with the homeland, in particular among the men, and the distinctive identity of the migrants in Britain, was boosted by a self-definition that highlighted contrasts with the values held by the host society – especially with the 'liberal' values of the 'West'. Their relative self-isolation made it easy to stereotype these Islamic groups summarily as a 'seed bed of terrorism'.

However, as Modood (2002) and Werbner (2002) note, a tendency towards politically motivated extremism is more common among educated young men from more privileged backgrounds than among members of the underclass. These young men are rather like the idealists of utopian movements, such as the anti-globalisation protesters, than the victims of racism, social disadvantage and everyday violence (Modood 2002). Tariq Modood also emphasises repeatedly that the possible involvement of a small number of British Muslims in acts of international terrorism must not be allowed to deflect attention from the need for the democratic inclusion of British Muslims as a social group, any more than the existence of the IRA ought to have been a reason to neglect or curtail the civil rights of Catholics in Northern Ireland. This comparative reference to a regional, that is, non-immigrant minority draws attention to the problematic indicated at the beginning of the previous chapter, where I used a well-known anecdote from Central Europe to illustrate that human beings do not need to migrate to find themselves on the 'wrong' side of a territorial boundary.

Culture and language between citizenship and identity

Although they belong to one of the oldest nation states, the inhabitants of the UK only became citizens by law as late as 1948. Before then, the inhabitants of the British Isles (in their entirety, that is, including the Irish[6] Free State) and of the British Empire were simply subjects of the Crown. Prior to the 'Nationality Act' of 1948, the 'British Nationality and Aliens Act' of 1914 related not to any 'British' territory but was directed at those who owed loyalty to the Crown. Until the 1960s all inhabitants of the Commonwealth of Nations as well as the remaining colonies were, accordingly, 'British subjects' with unrestricted rights to enter the British 'homeland'. As Commonwealth countries and colonies started becoming independent, these rights were increasingly curtailed. In the new 'British Nationality Act' of 1981, five categories of British nationality were introduced, only one of which – 'British citizens' – legally entitles the bearer to permanent residence in the country of his or her nationality. The other four categories are 'British Overseas citizens', 'Dependent Territories citizens', 'British subjects' and 'British Protected Persons'. A further category of 'British nationals (Overseas)' was added in 1987 to designate 'British' inhabitants of Hong Kong who did not have the right to enter the UK. However, exemptions were made for the 'patrials' – that is, 'Whites' from the former Commonwealth countries – who were free to migrate to the 'homeland' (Commission 2000: 206).

The introduction of citizenship in 1948 came as a reaction to pressures from the newly independent states, such as India, Canada and the Republic of Ireland, to define citizenship for their own respective jurisdiction. Consequently, a separate legal status was required for the remaining population of the UK. The Labour Party at the time supported what was referred to as the 'traditional', that is, the non-ethnic definition of 'British', as it had been developed as a political ideology since the union of the Crowns of England and Scotland in 1603. Against that background, it is clear why the Labour Party, well into the 1980s, regarded the nationalist movements in Scotland and Wales as a betrayal of the multi-ethnic identity of the 'British people'. The Conservative Party outwardly preferred a definition via ethnic self-ascription but with the 'British Nationality Act' of 1981 privileged the *jus sanguinis* over the *jus soli* (McCrone 2001: 101).

Even without the new immigrants, up to seven nationalities of 'indigenous' origin can be identified in the British Isles overall (Taylor 2001: 128), although it should be noted that all of these identities derive from groups that immigrated to these islands at some stage over the past two to three millennia. 'Rooted' in Great Britain, there is a southern or 'true' English nationality and a northern English one, along with Cornish, Welsh and Scottish nationalities; to these can be added an Ulster-Unionist nationality and an Irish-Nationalist one. For almost 400 years, an ideal-type 'British' identity, propagated initially by the Scottish king James VI after he was crowned King of England as James I, served as a kind of bracket holding these 'nations' together and thus providing the foundation for the functioning of a British polity. However, the term 'British' has lost much of its identity value and integrative power in recent years, especially among the younger 'white' population. The numerous Irish migrants living in Britain have long resisted the label 'British', a resistance that led to the adoption of 'Irish' as a separate ethnic category in recent population censuses. In Scotland, too, and not just since the partial devolution of the late 1990s, the designation 'British' has increasingly given way to 'Scottish', to the extent that it is nowadays almost easier to be British and Pakistani than to be British and Scottish (Modood 2001: 74). In this context, Krishan Kumar (2001) points out that England itself is no longer protected by an identity cushion of 'Britishness'. The flag of St George has increasingly replaced the Union Jack in recent years. Internationally recognised as an emblem used by many soccer hooligans, it has become, at the same time, a symbol for new forms of ethnic inclusion (Pines 2001: 57), signalling a tendential levelling of power relationships in ways that would be inconceivable with the imperial Union Jack. Similarly, the Union Jack is very strongly associated, especially in Ireland and western Scotland, with ethnic discrimination. While in Scotland the flag of St Andrew symbolises hope for the country's future, which includes immigrants (Khan 2001), in Northern Ireland the same flag has been used over the past two decades or so as a symbol of Ulster-Scots identity, expressing an affinity that includes a simultaneous distancing from the other components previously absorbed into 'Britishness', in particular 'Englishness'. In the past, compared to the British state as a whole, there have been considerably fewer immigrants in Scotland, Wales and both parts of Ireland, and indeed in rural England – both in absolute numbers and in relative terms – than in the major conurbations of England. This situation is changing slowly but steadily. The newly nascent national consciousness in the regions is therefore frequently seen as an opportunity to create a new, inclusive social order. In that regard, it is interesting to consider a little-known passage of the new citizenship legislation, according to which applicants for British citizenship must demonstrate a certain level of competence in the national language. Most commentaries[7] focus on English, which, on the one hand, is understandable, but reveals, on the other hand, a certain cultural bias on the part of the commentators, as the draft legislation stipulated 'English, Welsh or Scottish Gaelic' and hence treated the three 'indigenous' languages of Great Britain as of equal status for its purposes.[8]

However, even Home Secretary David Blunkett at the time showed little awareness of the matter. In a controversial essay on citizenship (Blunkett 2002), he demanded in conclusion – and without much connection with the rest of the text – that immigrants ought to learn English. Praise came, inter alia, from the Agency for Culture and Change Management, who particularly emphasised the need for women, in their role as mothers, to acquire a good knowledge of English. Keith Vaz, a Labour Party MP of Asian background, critisised the ethnocentrism of the essay and Ali Usman (2002) in a commentary in the *Khaleeji Times* pointed out that, especially in Northern England, the British National Party (BNP) constituted a serious threat for the Labour Party and that Blunkett was perhaps trying to address their supporters with his demands. Usman's warning appears to have been justified, as local elections in May 2003 showed, when the BNP increased its number of seats in the region from two to 13 – mostly at the expense of 'New Labour'. Many commentaries highlighted the virtual monolingualism of Britain and emphasised the value of multilingualism. Critics from Wales recalled that the Labour Party had always rejected as racist any demands that immigrants ought to learn Welsh.[9] With regard to this debate, it is also important to note that the home secretary dared to interfere in the private sphere of the citizens. This was widely frowned upon as 'un-British', but more significantly, it indicated an implicit assumption that the family is where successful socialisation of good citizens takes place, carried out by mothers who thus reproduce the nation; poor socialisation by mothers with a limited command of English leads, according to this logic, to vandalism and race riots (Werbner 2002).

Even leaving the immigration problematic aside, the meaning of 'British' identity is no longer self-evident. From 1997 onwards, 'New Labour' attempted vigorously to replace the traditional concept of 'Britishness' with a notion of 'cool Britannia' – progressive, open for change, culturally diverse and cosmopolitan. At one point the government's advisor on citizenship matters, Professor Bernard Crick, suggested that becoming British might be simply a matter of living in Great Britain and treating one another as 'British' (Appleton 2002). 'New Labour' evidently found it difficult to deal with the uncertain future of the state resulting from contemporary polycultural developments. On the one hand, there is nostalgia for the lost global empire; on the other hand, there are the contradictions within the liberal model of social inclusion and the attempt to develop a social democratic model of national economic growth in an increasingly globalised economic system.

Traditionally, ethnic identity has been expressed in particular cultural practices, and that is still the case. In addition, there is, now, also what Modood (2001: 72) calls 'associational identity' – an identity grounded not in descent but in social group labels and, in some instances, political activity. As the largest immigrant group in Britain, the Irish may be regarded almost as paradigmatic in this regard. Pushed equally towards assimilation and differentiation by their – imagined or actual – cultural proximity to the host society, this group has not only developed an 'Irish-in-Britain' identity but also includes today numerous members whose ethnic self-ascription rests on very shaky foundations even by postmodern standards. The urge to declare oneself as 'Irish' in Britain – and thereby as a member of a disadvantaged, marginal group in British society – has often less to do with ethnic origin than with an attempt to express political dissent while avoiding party political or otherwise ideological associations. The protagonists of such ethnic self-ascription deny that this turns 'Irishness' into an ideology-laden discourse, just as Marxists used to deny the metaphysical foundations of Historical

Materialism. Ideology is always something that 'the others' have, and so it is not unlike ethnicity, seen from the perspective of civic society.

Comparative perspectives

Due to its history of being composed of different, unequal 'nations', the UK is used to nuanced identities (Modood 1992: 24). In the contemporary debate on citizenship it is therefore often described as 'multicultural pluralist', in contrast to Germany, which has been characterised as 'ethnocultural exclusionist' (Koopmans and Statham1998: 691). At the same time, it is recognised (e.g. Klusmeyer 2001: 529) that even in the case of Germany we are dealing with a polity founded on cultural diversity and which can look back on a long history of more or less successful compromises between diversity and cohesion (see also Chapter 4). From this perspective, the main conservative party, the Christian Democratic Union (CDU), developed its controversial position paper on the idea of *Leitkultur* (guiding culture).

This idea of *Leitkultur* can be seen as the expression of a tendency among political élites in modern liberal-democratic states, whereby culture and other markers of distinction are used as a pretext for withholding full and equal membership of a polity from immigrants and other marginal groups (Klusmeyer 2001: 519). In this sense the idea was neither new nor unusual and certainly comparable to David Blunkett's approach. However, the debate in Germany and beyond acquired a special dimension through the adoption of the very term *Leitkultur*, which to many had resonances of older ideas of cultural superiority. Among the elements of *Leitkultur*, as they were laid out in the position paper, were the acquisition of the German language, a confession of loyalty to the German nation and recognition of the country's legal and political institutions (Klusmeyer 2001: 521). With these elements, the concept of *Leitkultur* is clearly inspired by the naturalisation rituals of the US. Whether or not the term may therefore be seen as an expression of political globalisation (McLeitkultur?), it may be noted that the development of the concept was an attempt to move from a negatively regarded ethnic definition of identity towards a positively regarded civil identity. Next to the US, France is considered a prime example of that approach (Nic Craith 2004b).

Since the French Revolution, the French state expects the diverse social and cultural groups in the country to surrender their local and group identities rather than trying to preserve them in a multicultural context: Basques, Bretons and other regional ethnic groups are simply French (DeCouflé 1992). That ideal of integration with its emphasis on a common French system of values and a uniform notion of what may pass as 'French' leaves little room for the maintenance of ethnic identities. The idea that anyone may simply become 'French' by adopting the national cultural model includes an expectation that ethnic identities in the regions as well as among any migrant groups will disappear with time (Kivisto 2002: 179). This expectation is based on the view of integration as an individual act, a view that leaves little room for group identities.

The ideal of an identity that is based on belonging to a political state rather than on ethnic descent can be understood, following Habermas, as *Verfassungspatriotismus* (constitutional patriotism). In post-Revolution France and in the US as a society of mass immigration, this ideal could gain a foothold sooner than in states that sought to define themselves via political continuity (notwithstanding any actual ruptures) and ethnic homogeneity (notwithstanding any actual diversity). In the UK, one possible version of *Verfassungspatriotismus* is grounded in loyalty not towards the legal and political institutions of the civic state but towards the Crown. Unionists and loyalists in Northern Ireland continue to hold on to this version (Kockel 1999a), while most of the other inhabitants of the UK found British citizenship as defined in 1948 a sufficient alternative – even if this was merely, inevitably, a kind of 'leftover identity' after the end of the Empire. In the second half of the twentieth century, the foundations of a civic British identity have been gradually eroded as the basic premises of a United Kingdom have been challenged. Such an identity continues to exist, for the time being, merely as an ever-weakening bracket holding together a range of other identities – including resurgent civic identities in Scotland and Wales and the uneven reassertion of English regional identities.

Klusmeyer (2001: 522) points out that the CDU position paper clearly foregrounds the cultural implications of immigration in its preamble, where it states that immigration policy and integration policy can only succeed for people assured of their own national and cultural identity, founded on a cosmopolitan patriotism (cf. Christlich-Demokratische Union 2000). In this commitment to the principle of establishing an assured identity for the majority population, the ideologies of the CDU and 'New Labour' converge. It remains questionable for how long notions of cultural homogeneity can be maintained in the face of rapid economic globalisation and transnational media and communication networks.

The holistic vision of a national culture obscures the differences between its individual components and can thus throw into sharper relief the lines of separation between 'insider' and 'outsider' (Klusmeyer 2001: 525). But that exclusion is by no means inevitable. The CDU position paper presented the concept of a *Leitkultur* as a 'third way' between French-style assimilation on the one hand and, on the other hand, multicultural segregation leading to the creation of US-style ghettos. From that perspective, the multicultural model is characterised by the CDU as *unverbundenes Nebeneinander* (unconnected side-by-side existence) of human beings living in parallel societies.

Klusmeyer attributes this evaluation of multiculturalism to a conservative world view. With regard to the British situation, the same view of multiculturalism is shared by immigrants. For example, Tariq Modood (1992, 2002) is critical of a 'politically correct' multiculturalism that promotes cultural fragmentation rather than contributing towards a more integrative 'Britishness'. Yasmin Alibhai-Brown (2000) detects in multiculturalism the danger that it might raise up barriers between the 'tribes' making up British society today, instead of promoting cultural and political cohesion. Many regard multiculturalism as, in Alibhai-Brown's words, 'something Black people do', and it is understandable if English people feel neglected because their ethnicity appears to be denied while all the other identities present in the Kingdom – be they Welsh, Hindu or whatever – seem to be celebrated. Instead of the multicultural alternative – complete assimilation or segregated parallel societies – Alibhai-Brown proposes parity of esteem between the different groups. The experience of the 'peace process' in Northern Ireland shows that this is by no means an unproblematic solution (see, e.g. Nic Craith 2002, 2003). In the German context, Klusmeyer (2001: 526) notes that for marginalised minorities the legal recognition of formal freedoms on its own is not sufficient to level the imbalances of power that determine the opportunities for individuals and groups to participate in society. In the post-1989 Federal Republic of Germany, the granting of (Federal) German citizenship to the population of former German Democratic Republic or to ethnic Germans from Eastern Europe was by no means enough to ensure their effective integration (Klusmeyer 2001: 528).

Evaluation

Immigration always raises complex issues of cultural adaptation. Attempts to reduce the challenge of integration to aspects of a shared cultural identity neglect the issue of power. When any political élite insists on treating this challenge primarily as a matter of cultural conformism, one might well ask whether the protagonists are perhaps more interested in creating obedient subjects than in promoting democratic civic virtues (Klusmeyer 2001: 528). Any calls for cultural adaptation, for example through the learning of a language, should primarily serve, as Klusmeyer argues, to facilitate participation in the host society – not the fulfilment of some prescribed collective identity.

Kymlicka (1995) has described polities comprised of voluntary migrants as 'polyethnic states' where the question of territory remains open because the new arrivals cannot raise territorial claims. To the extent that the UK is able to resolve the question of territory by a gradual disintegration into its constituting elements, the British Isles may see the formation of polyethnic polities in which the core questions will be how the various groups wish to assimilate (and who with) and in how far the majority population is prepared to take part in this process.

In Northern Ireland, which has experienced much less immigration from outside these islands than other parts of the UK, the territorial issues discussed in the previous chapter are likely to persist for a little longer than elsewhere. Even there, the encrusted two-cultures model that is still used to characterise the region, on the ground as well as in academic analyses, has been shown as inappropriate for some time (Nic Craith 2002) and is becoming more so as migration flows are changing the ethnic composition of the regional society (McDermott 2008).

If, following Alibhai-Brown's (2000) vision of a society founded on parity of esteem, the UK changed from a polity where cultures exist side by side, to become a polity where they genuinely exist together, then that would be the first step away from the multicultural dilemma where the choice is either complete assimilation or mutual separation and towards a networked polycultural society. I see a key contrast between a multicultural society and a polycultural one in the latter being polycentric rather than centralised and thus less hierarchical than the former. A polycultural society therefore creates a societal whole in which all can participate equitably (if they so wish). This does not solve the power question entirely, but the power of definition at least lies with the many different centres rather than with one or a select few.

That may sound utopian, and a recipe for anarchy. A decade ago the Parekh-Report (Commission 2000) noted that the immigration debate in the UK has for a long time seen immigration as a problem rather than an opportunity. The events of '9/11' and the subsequently declared 'war on terror' have ensured that this evaluation has barely changed. Meanwhile, however, reports from other European countries concerning looming shortages of skilled labour have also had some impact on British policy thinking. An added complication is the spectre of disintegration – for better or worse – of the UK as a political vision. Whether a different party in government is able to tackle these issues better than 'New Labour', with its monocultural vision of a 'cool Britannia', has done over the past 13 years remains to be seen. As I am writing this, the BNP is recording gains in European parliament and Westminster by-elections that may signal any one of several possible scenarios.

Lithuanians in Northern Ireland

The debate on immigration and integration tends to focus on non-European migrants and largely ignores groups that are considered 'invisible' on grounds of their outward appearance. That such distinctions are ill-founded is easily demonstrated. Many years ago during fieldwork in the West of Ireland, I used to play a game with a local man who worked in an independent hostel. We would sit in the corner of the lounge and predict the nationality of new arrivals, taking our clues from simple indicators such as dress and mannerisms. Four times out of five, our guesses proved correct. To reverse the example: I have lived in different parts of these islands for more than 25 years. While shopping one Saturday not so long ago, my wife and I were approached by a stranger with the observation that we did not look as if we were 'from here'. There is no such thing as an 'invisible migrant'.

I began this chapter with reflections on one allegedly 'invisible' migrant group, the Irish in Britain and on the continent, and continued with another such group, the Germans in these islands. Before moving on to the next chapter, I want to take a brief look at a third such group, because their experience raises questions that have not come up yet. My discussion of this group, Lithuanian migrants in Ireland, draws on a recent doctoral dissertation at Vytautas Magnus University, Kaunas (Liubinienė 2009), as well as on my own observations.[10]

According to statistical sources for 2004, Poles and Lithuanians formed the largest groups of immigrants from the new EU member states in the UK (Office for National Statistics 2006). As immigrants from new EU member states are mostly economic migrants, their impact in the economic sphere has attracted significant interest and is well acknowledged. The social and cultural spheres, while of equal importance, are largely neglected in current research. In the Republic of Ireland, the Equality Authority has published a report (Conroy and Brennan 2003) on migrant experiences based on a combination of methods and addressing some cultural issues.

For Northern Ireland, which has attracted a disproportionately higher number of migrants from new member states than other parts of the UK (Jarman 2006: 50), a preliminary survey of migrant workers has been carried out (Bell et al. 2004). While Polish migrants are numerically the largest group entering the UK and Ireland from the new EU member states, Lithuanian migration to Northern Ireland is comparatively significant, especially when set in the context of the sending society (Table 3.3).

A comparison of official figures is fraught with certain difficulties inherent in the data collection process – including differences in classification between different sources – and influenced by factors such as non-registration of immigrant family members. Even so, the comparison raises interesting questions not just in terms of possible differences in migrant motivations and pre-migrations perception of the host country but especially concerning the cultural impact of emerging links between the two societies. While the Lithuanian population is only 8.8 per cent of the population of Poland, Lithuanian immigration to Northern Ireland stands at 41.5 per cent of Polish immigration and in the Republic of Ireland it is even higher, at 44.1 per cent. Moreover, in statistical terms, nearly 1.5 in every 1000 Lithuanians have migrated to Northern Ireland and more than 5 to the Republic of Ireland compared to just over 0.3 and 1, respectively, in every 1000 Poles. In other words, in terms of statistical relativity, five times more Lithuanians than Poles are migrating to Ireland – which raises the question: why? National Insurance Number registrations suggest that while Northern Ireland is

Table 3.3   Comparison of Lithuanian and Polish migration to Northern Ireland

	Population	Northern Ireland (1)	Ratio 1/000	Republic of Ireland (2)	Ratio 1/000	National Insurance No. Applications 2004/5 (3)	
						% of NI	% of UK
A. Lithuania	3.4m (2006)	4,987	1.46	18,063	5.31	23	14
B. Poland	38.6m (2005)	12,020	0.31	40,973	1.06	47	57
Ratio A/B	8.8%	41.5%		44.1%			

Sources: (1) Jarman (2006: 48)
(2) Department of Social and Family Affairs, Dublin (2005)
(3) Jarman (2005: 9)

less popular with Poles than other parts of the UK, with Lithuanians it is considerably more so. What these statistics indicate is that a growing proportion of Lithuanians 'at home' have family links with the British Isles – in particular with Northern Ireland and the Irish Republic. What makes the island such a popular destination and how do these growing links shape the culture of both parts of Europe? Given the population ratios in Table 3.3, one could expect a much stronger cultural impact of Ireland (both North and South) on Lithuania than on Poland. What kind of cultural links are emerging between the two societies as a large and growing number of Lithuanians 'at home' will have family connections in Ireland?

Liubinienė (2009) uses the concept of *savos erdvės* (own space) to analyse the often contradictory ways in which migrants appropriate and structure the places and spaces that make up their worlds, in the process creating new, mobile and yet deeply rooted identity sets. Similarly, my research on German migrants suggests that new identity sets are emerging, fostering links with the country of origin that are less conspicuous in terms of outward cultural expressions but structurally as deep, if not indeed deeper, in terms of kinship and social networks, than in previous migrant generations. To investigate further this apparent contradiction through comparative analyses of a range of European migrant groups, and – with 'ethnic' hostility increasingly affecting European 'internal' migrants in many countries – to explore paths towards a culturally more open-minded society, would be key aims of the envisaged research programme.

It could be argued that the cultural interpenetration that occurs in the ethnic frontier is a key process of progressive 'Europeanisation' that creates a 'people's Europe', a kind of 'unity in diversity', to cite just some of the catchphrases commonly used in EU literature aimed at informing citizens about European integration and securing their support for the venture. However, among the new member states in Central and Eastern Europe there is a growing sense that Europe, and its actual and potential Europeans, may be lost as the process of integration unfolds.

The lost Europe and its Europeans

The idea of Europeanisation is closely linked to the notion of a Europe without borders, and much of EU policy and practice is concerned with minimising, or indeed eradicating the impact of borders as barriers to the free movement of factors of production. Whether a Europe without borders can ever become a Europe without boundaries is a question that would deserve further investigation. Borders, the physical and politico-legal expressions of boundaries, may well be removed to facilitate the free flow of goods, services, ideas and labour, but the maintenance of the underlying boundaries will remain an imperative for some time yet. Much ink has been spilled in the final decades of the twentieth century about the end of the nation state, but even that rhetoric has stopped at the idea of a 'Europe of the regions', thereby recognising that, as we shift the boundaries of the state 'upwards' and 'outwards' to create a 'Fortress Europe', and simultaneously 'downwards' and 'inwards' to create stronger regions, we are not transcending the nation state as the conceptual foundation of territorial polity, but merely reshaping it. With regard to the ideal of a 'Europe of the regions' that has enjoyed increasing popularity at the political level, it must be noted that whereas 'region' implies differentiation and diversity, labour mobility according to market principles works more smoothly if the labour factor is, and becomes, more uniform and pliable. Proponents of a free-wheeling postmodernity with its cheerful identity warehouse should be aware that their vision ultimately plays into the hands of those who wish to move labour around the unbounded market only according to market signals.

However, are ethnologists and other students of culture and society becoming overly concerned with migrants and their mobility when the vast majority of people still die within ten miles of their place of birth? Do our studies of migration need to focus more on the sending society than, as they have conventionally done, on the host society? As well as, or perhaps even instead of, looking at those ethnic frontiers created by inward migration, should we focus on how Europe is lived and built (or destroyed) by those who stay behind but, in some cases at least, have ever-expanding links abroad?

Whichever end of the migration process we focus on, when we talk about Europe in these terms as a growing ethnic frontier where cultural encounter and interpenetration takes place, we have to face the question of what happens to Europe and its Europeans. Where is this Europe, is there a Europe at all, and if so, what does it look like?

These concerns find expression in popular and, increasingly also, academic literature and in panels at international scholarly conferences, such as the one that I co-chaired, with Rajko Muršič from the University of Ljubljana, at the congress of the European Association of Social Anthropologists in Bristol in 2006. In Chapter 6, I explore this lost Europe and the potential role of the ethnologist/anthropologist in recovering it, but first I want to take a closer look at the marketplace where mobile Europeans are supposed to trade in their futures.

4
Third Journey – To the Market:
Trading Our Futures

Some 15 years after the collapse of Communism, the eastward expansion of the EU brought many formerly Communist countries into the Common Market. In Lithuania, the eastern boundary of that market runs about 30 kilometres east of Vilnius – approximately the same distance from the city as the geographical centre of Europe, which is marked some 25 kilometres to the north of Vilnius.[1]

Since the discovery of Europe as a research field for cultural anthropology, Eastern Europe in particular has taken over the place of the exotic 'other' region within or at home. In a way, this reflects the view expressed some time ago in the *Financial Times* (8 May 1993) of Eastern Europe being 'a distant part of the world ruled by medieval passions, which are the antithesis of everything modern man stands for'. However, it is not so much the 'medieval passions' that have attracted the interest of anthropologists but rather the process of transition towards a capitalist system. At the very heart of the capitalist system is the concept of 'the economy' as an unconstrained sphere, an invisible actor working in accordance with its own rules. Verdery (1997: 716) observed that the restoration of private property in Eastern Europe has brought the 'complexity of the very idea and possible forms of property ... more fully into view', raising the question of how 'this notion of possessing an exclusive right that is so central to Western selves and Western capitalist forms' is actually culturally constituted. The reintroduction of private property, together with the renewed commodification of land and labour, mark fundamental departure points for 'theoretical critiques of ... commodification and its place in social science theory'.

European ethnologists should be able to make a significant contribution to this critique, both in the East and in the West, and this not just because our field has some significant roots in political economy

(Kockel 2002a). The New Economy has become 'something of a grand narrative, surrounded by buzzwords like the network society, globalization ... or "the knowledge society"' (Löfgren 2001: 155). Orvar Löfgren argues (op. cit.: 159) that

> the social and cultural organization of imagination, dreams and fantasies is a very underdeveloped field in the European ethnology. ... A historical perspective could show us how the ... New Economy differs from earlier dreamworlds. What happens, when you try to institutionalize, standardize or commoditize fantasies and other products of the imagination?

He suggests a return to the 'classic tradition' of the 'folklorist interest in the world of fantasies and dreamworlds, its genres, imaginary and social contexts', starting with what he calls an 'archaeology' of the rhetoric of a New Economy with its 'extremely ahistorical ... understanding of the world'.

Such a return to the origins of our field should also include a critical engagement with the disciplinary past that European ethnology has not yet had, by turning the spotlight onto its roots in *Gesamte Staatswissenschaften*[2] (general political science), where it evolved once in close proximity to the Historical School of economic thought. Political expediency pushed the discipline of economics in a different direction during the nineteenth and much of the twentieth century. The Historical School matured into the concept of a social market economy, which offered a genuine Third Way, not just a thinly veiled attempt to introduce neoliberalism by the back door. While it certainly did not remain immune to temptations of imperialism, the *Nationalökonomie* of the Historical School – in both its 'older' and its 'younger' politico-epistemological incarnation – continued to be shaped by its cultural and geopolitical origins in Central, and especially German speaking, Europe rather than in the transatlantic realm of global empires. Conscious of its common roots with a historically thinking political economy, European ethnology as an empirical approach to the study of everyday culture could, and should start to ask some politically inconvenient questions. In doing so, it would be in good company: economic anthropologists (e.g. Gudeman 2008) have long emphasised the 'economic tension' at the heart of the real life economy, which depends on the mutual contingency of market and community. By contrast, a new 'virtualism' (Carrier and Miller 1998) is progressively compelling everyday lives to conform to the requirements of economic thought in much the same way as the Medieval Church once developed its own complex eschatology as a normative framework that would guide everyday praxis. The 'hectic frenzy surrounding life in the New Economy calls for a critical, historical and reflective perspective', says Löfgren (2001: 162) and wonders whether 'the important social and cultural changes are occurring on other levels or in different areas from those which the media and much of the current research are focusing on'. Considering how the current economic crisis seems to have caught the established economic powers in business, politics and academia by surprise, one may well conclude (see, e.g. Schwegler 2009) that some key places and practices must indeed have escaped the attention of the neoliberalist hegemony.

In this chapter I look at economic issues through a different lens. A detailed exposition of an alternative framework will have to wait for another occasion; what I want to do here is to explore some historical and experiential foundations, together with basic premises and presuppositions, in the context of my search for Europe. After looking at the geopolitical framework that may yet generate a third or fourth generation of the Historical School, I first sketch the outline of a theoretical perspective grounded in such a framework, and then offer some reflections on market rationality and fundamentalism, prompted by the changing East–West relations in Europe.

Where is the German nation?

From its beginnings in *Allgemeine Statistik* through its sociological inflection by Max Weber to the Christian socialist vision of the ordo-liberalists and the Third Way advocated by Ota Šik at the time of the Prague Spring, the *Nationalökonomie* developed in the tradition of the Historical School has been deeply rooted in the history and politics of German-speaking Europe and its geopolitical and cultural sphere of influence. As indicated in the previous chapter, the German polity is founded on historical diversity and can look back on a history of more or less successful compromises between this diversity and cohesion. The search for a German identity has exercised some of the finest minds in that part of Europe since the eighteenth century, without any noteworthy consensus being reached to date (Bausinger 2000; Gelfert 2005; Fuhr 2007). Following the unification of two somewhat German states in 1990, the neighbours in Europe are watching with concern any stirrings of nationalism from a region which has procured rather devastating versions of this ideology over the past two centuries.

Gudrun Tempel, a German woman who emigrated to England in the 1950s, wrote a book in 1962 under the title *Deutschland? Aber wo liegt es?* – Germany, but where is it? She had a point. Where on earth is Germany? And what makes Germany, or anywhere else, for that matter, well – German? Having lived outside the boundaries of the political entity called Germany for over a quarter century, I know more about it now than before I left. Because of the atrocities committed in the name of our presumed nation, being born as a 'German national' had no positive meaning for most of my generation, who on their youthful forays into other parts of Europe, as InterRailers or exchange students in the 1970s, had often been greeted with Nazi salutes and sometimes challenged to explain the unexplainable. The only significance which being German had – beyond this – was that one belonged to a society where even the poorer classes seemed economically better off than some of the well-to-do in neighbouring countries. Our identity was not constructed around the concept of a German nation, which existed only as a utopia for certain people who seemed to fill the history books with abortive attempts to achieve it. But, contrary to the perception of outsiders who tried to analyse our national soul, this did not mean that we had no identity or, worse still, were suffering from an identity crisis. The big mistake these commentators made was that they proceeded from the wrong premise, namely that, as in France or Britain, there is an actual basis for a national identity in Germany. To my mind, the question of a German identity is entirely misplaced because – whatever the pundits on the rostrum of international politics (or, for that matter, any Neo-Nazi propagandists) might like to make us believe – Germany is not a singular nation. If anything in this line, it is an association of nations, a multinational federation. More to the point, I think the category of the 'nation' is inappropriate for an analysis of identity in the German context. In the remainder of this section, I try to explain why and suggest an alternative framework.

The *Wirtschaftswunder* as glue of the nation

In West Germany during the early decades after the Second World War, the question of a German national identity, for a long time, seemed to be important only in the context of the *Grundgesetz* (constitution; lit.: Basic Law), which postulated the integrity of the national territory without specifying too clearly the eastern boundaries of the same. The only aberration from this focus was the debate on 'the Left and the National Question', which surfaced during the students' unrest in 1967/8 and occasionally again during the 1970s. However, since the

Historikerstreit some years ago – a lengthy and vigorous dispute over revisionist historiographies of National Socialism – much ink has been spilt in a renewed attempt to come to grips with the vexed question of nationhood.

Harold James (1991) suggests that German historians had been surprised by events in Eastern Europe because they had excluded the question of the historicity of nations from their inquiry and many had simply substituted 'society' for 'nation' as an analytic concept. James argues that German national sentiment, ever since the *Zollverein* (customs union) of the 1830s, and most certainly after the declaration of the second *Reich* by Bismarck in 1871, has been dominated by economic ideas and expectations. Right across the political spectrum, and in all walks of life, James detects this equation of *Nationalökonomie* (national economy) with *Nationalstaat* (nation state). He sees this economistic orientation as a major cause of the pessimistic critique of culture which became so fashionable in the early decades of the last century, generating the growth of populist movements of which the Nazis soon became the most influential one. After 1945, the idea of 'nation' (together with other concepts, such as *Volk* or 'tradition') was tainted by the use the Nazis had made of it and thus, according to James, neither West nor East Germany developed into a nation proper. To be German was to be a member of a society that had created the *Wirtschaftswunder* (economic miracle) after the war. This applied equally on both sides of the Iron Curtain. One was proud of the economic achievements of one's state, and this pride was sufficient to provide a sense of belonging.

In his analysis of the importance of economic concerns, James is quite correct. The identification of the post-war generations with the *Wirtschaftswunder* and the policy of the coalition government under Helmut Kohl towards both German unification and European integration are ample evidence for the persistence of an equation between the economy and the nation: if only the accounts work out alright, the German nation will be recreated without any problem and a European identity will emerge from a common currency. And yet James is wrong when he asserts that it has been this emphasis on the economic which prevented Germany and the Germans from developing a proper national identity. He is as wrong as the German liberals in the nineteenth century with their essentially – if subconsciously – imperialist frame of mind. A German identity, much more so than a British one, and quite unlike a French one, can only be supranational. There is, in effect, no German nation.

A grass-roots perspective on 'German' history

German history books commonly date the origins of 'Germany' to Emperor Otto in the tenth century. However, as most history is written retrospectively, for Otto and his contemporaries the term *Deutschland* is unlikely to have had the meaning it acquired during the rise of the nation state from the late eighteenth century onwards. The Holy Roman Empire, as the first German *Reich* was commonly known, was not a unified entity, but a ragbag of more or less autonomous territories, free states, imperial and free cities over which an Emperor exercised some measure of authority in a rather limited range of political and legal matters, and by no means unchallenged. For a while, this first *Reich* was, just as Britain and Ireland were at the time, a playground for the territorial aspirations of Norman barons. In later centuries, the autonomous polity of the Teutonic Knights, the persistent defiance of imperial powers by the Hanseatic League of free cities or the power of the Fugger family, who treated the Emperor like a puppet on a string, are well documented. Through the Norman connection with the Mezzogiorno, and later through the Fuggers, the 'core' of the *Reich* was oriented towards the Mediterranean, in particular towards Italy and, towards the end of the *Reich*, Spain. Latin, and later Italian, was the language of trade and, to a large extent, politics at the 'international' level. The northern part of the *Reich* looked towards Scandinavia and, again through Norman connections, to the lowlands of Scotland. The lingua franca here for many centuries was Low German and it continued in this status, albeit somewhat diminished, even after the decline of the Hanseatic League in the seventeenth century.

This division of the *Reich* was by no means incidental and it has repercussions to the present day. The dividing line runs roughly along the northern boundary of the ancient Roman Empire. North of this line, the western part of Germany was settled by Celts, the area east of the river Elbe by Slavs and only the wedge in the middle, which later, under the Saxon kings, expanded in southeast direction, was Germanic. The colonisation of what are now the five eastern states of Germany, and of the territories beyond, was undertaken about the same time as the Norman conquest of Wales and Ireland. However, while the East/West cultural divide retained some significance even into this century, the real regional divide is between the north and the south. Expressions of it can be found in many spheres of cultural life. Historically, enshrined in the 1648 Peace of Westphalia, the south is predominantly Catholic, whereas the north of Germany is mostly Protestant. The south has traditionally tended to vote Conservative, the north more Social Democratic.

Dialects of Middle and Upper German, as closely related as Irish and Scots Gaelic, are spoken in the south, and it is from one of these dialects that contemporary 'High' German derives. In the north, dialects of Low German, as distinct from 'High' German as Welsh is from Irish, are spoken alongside Scandinavian languages like Danish and Frisian. The list goes on and on, through marked differentials in wealth and employment, to social values, to life cycle rituals and geopolitical orientation (towards France in the south, Britain and the Nordic/Baltic states in the north) – to name but a few.

The quest for a German national identity is closely linked to the emergence of Prussia as the main hegemonial power among the German nations from the late eighteenth century onwards. Prussian imperial aspirations created the need for an overarching identity which would serve to hold the growing empire together. The people of other German nations were reluctant to be convinced of this common identity, and there is a wealth of ballads from the nineteenth century, such as Ludwig Pfau's *Badisches Wiegenlied* (Badensian Lullaby; Friz and Schmeckenbecher 1979: 268) of 1849, which give eloquent expression to what 'ordinary' people thought of being made 'German' more or less at gunpoint:

Der Preuß hat eine blut'ge Hand,	The Prussian has a bloody hand.
die streckt er über das Badische Land,	He stretches it out over Badensian land,
...	
Gott aber weiß, wie lange er geht,	God knows how long he'll go on like this,
bis daß die Freiheit aufersteht,	before freedom will rise up again,
und wo dein Vater liegt, mein Schatz,	and where your father is buried, my darling,
da hat noch mancher Preuße Platz!	there is yet room for many a Prussian!

And even late in the twentieth century, a radical Bavarian writer fantasises about what might have been had Bavaria, in the fateful war of 1866, defeated the expansionism of Prussia (Amery 1979), or, with a view to the wider, European scene, if the Vatican had a time machine enabling it to undo the political blunder that led to the Stuarts losing the crown of Britain and Ireland – a feat which would annihilate the Act of Succession and thus put a Catholic Bavarian prince or princess, descending in direct line from Charles I, on the throne of the UK, in the same process potentially doing away with this non-entity, 'Germany', in favour of multiple autonomous territories (Amery 1984). It is, of course, worth noting here that within the Federal Republic of Germany, founded in 1949, Bavaria remained a free state, pursuing its own policies in many areas – including in the international sphere.

A Germany of the nations and their regions

When the Federal Republic of Germany was established, it was organised as a federation of states who delegated certain powers to joint authorities. Among those areas where the states retained all powers was the entire sphere of culture, including education (*Kulturhoheit*, meaning the cultural sovereignty of the states). In accordance with Article 29 of the original Basic Law, the states were constituted along historical and cultural lines, an important principle being that of *landsmannschaftliche Verbundenheit*, that is, in plain English, national identities. In the territorial organisation of Germany, these are most clearly expressed in the *Regierungsbezirke*, the provinces of each state. *Bayern* (Bavaria), for example, comprises *Ober-* and *Niederbaiern*, *Schwaben*, *Ober-*, *Mittel-* and *Unterfranken* and the *Oberpfalz*, thus giving regional recognition to the Bavarians, Swabians, Franks and Palatines, who make up the indigenous population of *Bayern*. A *Landsmann*[3] is someone who comes from the same country as yourself, and with whom you share a cultural identity. Very few Germans, if asked to define themselves in terms of their *Landsmannschaft*, their national affiliation, would simply answer: *deutsch*. You might hear *fränkisch*, *hessisch*, or *friesisch*. These are, if there are any at all, the German nations. The question of a German national identity is, therefore, purely academic. My great-grandparents were born as *Franken*. There was no Germany then. By the time they were married, they had been 'German' for only five years. When I grew up, the world of my cousins consisted of Upper Franconia, that is, one of the northern provinces of Bavaria, and Thuringia across the Iron Curtain. West and East Germany only had a meaning in the context of this local border and Germany as a whole was, well, pretty irrelevant. When they talked of the capital city they meant Munich, not Bonn or Berlin.

But is this not simply regionalism, which we find in many nation states, even the smaller ones like Switzerland, where the 'Jura-question' led to the creation of a new canton? I would argue that it is not. The movement for a Franconian state, uniting the three Bavarian provinces with their 'lost territory' that belongs to Baden-Württemberg, is a case in point. The Franconians within Bavaria, whose 'real' capital is Nuremberg and who may desire greater autonomy for their three provinces within the state, are regionalists, as are their counterparts across the border. The discourse is over differentiation from the Bavarians and their state – Germany, once again, is irrelevant for the purpose of identity. The Franconian nation, according to the movement, is divided into regions which now belong to two different states when it should really have a state of its own. The Bavarian response to these demands, so far, has been that of the typical nation state towards its regions. Some years ago, the Bavarian government, in which the *Baiern* dominate over the *Franken, Schwaben* and *Pfälzer*, decided there was a need for new border poles and plaques along the border, not with Austria, Czechoslovakia or the German Democratic Republic but with the other states of the Federal Republic. This, the *Bundesland* (federal state), is the highest political level at which, in the context of 'Germany', we can usefully and sensibly speak of a national identity. Beyond it, we are all Europeans anyway. 'Germany', as such, only denotes those territories outside Austria and Switzerland where German is the first language of the majority of the population.

If we want to think of Germany in terms of 'nation', we can only look to the level of *landsmannschaftliche Verbundenheit*. In the literal sense of the term, 'nation' means the place, or region, where you were born and, in most cases, grew up. But the strength of identification in 'Germany' is with the ethnic region. The post-war refugees and expellees, although 'German' themselves, remained exiles in 'Germany'. To them, Pomerania, Masuria or Silesia are more meaningful concepts for the purpose of identity. Among a deluge of more or less autobiographical books published in recent years by people from these territories, one text stands out as an analytical treatment which transcends the mere reminiscences of a lost heritage. Krockow (1992), a Pomeranian by birth, introduces his book by explaining why he could never become a Saxon, although he spent most of his life in Lower Saxony and all the usual markers around which identity is constructed are rooted in that country. He then proceeds to explore the meaning of *Heimat*, which he calls 'a German theme'. Foreigners are always bemused by the Germans' apparent obsession with *Heimat* and by the insistence that the term cannot be accurately translated. Without, at this point, wanting to go into the philosophical intricacies of the concept, some of which are discussed later in this chapter, I would venture to suggest that *Heimat* is a more appropriate notion than 'nation' when we try to come to grips with identity in the German context. Territorially, *Heimat* is defined by ethnic regions, such as Franconia, or the Aleman cross-border region *Dreyeckland*, between 'Germany', Switzerland and France. Politically and ideologically, it is a task (cf. B. Schmidt 1994), unlike the 'nation', which is a given. Thus the conceptual framework of nationalism and the nation state is not particularly suitable for an analysis of 'national' identity in the German context. Certainly, attempts have been made throughout the last 200 years or so to construct a national German identity. These attempts have failed, not because – as James (1991) would have it – the Germans have been unduly obsessed with economic performance. The economic aspect has, in fact, been the only major unifying factor among the German nations and as such the only real basis for any German potential national identity.

In 1948, the philosopher and anthropologist Helmuth Plessner proposed a federation of German lands as the core of the new, post-war Europe. He argued that this would prevent the re-emergence of a powerful nation state in the centre of the continent that could once again drag the world into devastation. His proposal anticipated, after a fashion, the idea of a 'Europe of the regions' that gained currency some 40 years later in the process of Western Europe-based EU integration, at the time when the political system of 'the other Europe' beyond the Iron Curtain began to disintegrate. While regionalism had been on the horizon for some two decades, many political commentators had brushed it aside as an anachronism leading to 'the Balkanization of practically everyone' (Zwerin 1976), and this discourse continued into the 1990s (e.g. McFarlane 1994). As long as the primary rhetoric associated with – what was then still Western – European integration remained squarely economic, the emphasis was on civic values that favoured the construction of an orderly, structured but essentially unified polity, arguably modelled on the *citoyen* ideals of the French Revolution.

There has been a steady flow of arguments taking issue with the standardising force of that political esprit emphasising by contrast diversity as one, if not the, essential defining characteristic of Europe (see Kockel 1999a: 35–6). Usually such arguments are regarded, quite rightly, as having their roots in the Romantic Movement, which, because it also fed into National Socialism, is widely regarded as a discredited curiosity of post-Enlightenment Europe (see, e.g. Lepenies 2006). This is not the place to debate whether the Jacobin legacy is entirely without blemish as its protagonists suggest, but it is worth noting the transatlantic interpretation of this juxtaposition, neatly summed up by the former US defence secretary, Donald Rumsfeld, in terms of 'old Europe' versus 'new Europe' – the former being caught in a Romantic time warp and holding on to discredited, Romantic European values, the latter being progressive and, well, American in its world view. During the last two decades of the twentieth century, however, the idea of 'unity in diversity' enjoyed growing popularity at the European level and became a major slogan employed by the EU. Contemporaneous with the decline of Communism, this led to a growing discourse about grand and ultimately, as some (e.g. Lepenies 2006: 411) would say, unanswerable questions concerning a European identity and the boundaries of Europe.

Following on from my earlier explorations of the ethnic frontiers of European integration (Kockel 1999a), I suggested at the 2004 conference of the European Association of Social Anthropologists in Vienna that one possible answer to questions of European identities and boundaries may be to examine whether and to what extent Europe can be considered as a *Heimat* being built by people who, like the 'settlers' and other migrants considered in the previous two chapters, dwell in the frontier created by EU integration.

Finding one's place in a Europe of free-floating regions[4]

As already noted in the previous chapter, in the context of German migration to the British Isles, the concept of *Heimat* has been extensively and intensively debated for some time (Hartung 1991: 143–4), not only within European Ethnology, which has an affinity with the term that could be described, entirely without irony, as traditional. At this point I do not want to delve into the etymological derivation or the historical development of the term, nor will I examine the ideological content it acquired during certain periods and at specific junctures. Instead, I want to develop the concept of *Heimat* as a form of everyday resistance, building on earlier considerations concerning the juxtapositions of *Heimat* versus *Herrschaft* (hegemony; Kockel 1989b) and *Heimat* versus *Fremde* (Kockel 1999a). For the moment, I will accept 'the other' as a sufficient, if not entirely unproblematic translation of *Fremde*. These considerations of *Heimat* draw on field experience as (more or less participant) observer during fieldwork projects, most of which dealt – in one way or another – with questions of regional culture and economic development in Europe. Arising from that context are certain inevitable limitations of the perspective as well as certain foci; in particular, it should be noted at the outset that the economic, commonly treated as a sort of ubiquitous, background that explains everything – and therefore too much and nothing – is critically important in these considerations.

The re-emergence of a politicising discourse of *Heimat* since the 1960s has been closely connected with local and regional protest, reform and various other movements. Partly due to problems of categorisation in relation to the vocabulary used and the evaluation of the 'authorities' invoked by these movements, it has been difficult to determine in many cases whether a particular movement ought to be regarded as progressive, conservative or reactionary. The subject is fraught with highly emotional interpretations leading to easily triggered stereotypes. Moreover, the limitations of our analytical vocabulary make any discussion of the issues I am raising here uncomfortable to say the least. Nevertheless it is important that we do not try to discourse them out of the way by finding labels for them that permit us to 'other' them into convenient categorical boxes. Historical precedence should alert us that the lids of such boxes do not shut firmly but allow the political ferment inside to build up to explosive strength.

In the light of the contemporary debate on globalisation/glocalisation, progressive European integration and a rekindling of nationalism and racism, the matter of *Heimat* as a political issue needs to be theoretically reflected – with all due caution – more thoroughly today than ever before. It is no longer sufficient to declare *Heimat* as imaginary or invented, using a liberal sprinkling of (by now somewhat old-fashioned) references to Benedict Anderson or Eric Hobsbawm and to restrict the discussion to the question of how and for what purpose people have tried to construct stable virtual pasts for their memories to reside in. I come back to that issue in the next chapter. In the present section, I can only open up the debate, offering a perspective but no concrete conclusions, an invitation to consider carefully whether *Heimat* and related concepts can be thought of in different ways than the ones we are used to and to see whether on that basis a terminological apparatus may be developed that is worth the effort.

'Heimat' as a political issue and individual experience

Already in the arbitrary year of 1979, Ina-Maria Greverus identified *Heimat* as, among other things, a political task involving the active acquisition of an identity space through dwelling and individual creative appropriation (Greverus 1979). Thus the term was being detached from its traditional anchoring in the past and was instead posited within an active present. As a process in the contemporary force field of everyday, conscious appropriation of the world in confrontation between yesterday and tomorrow (cf. Hartung 1991: 146), we find the term *Heimat* in the work of the philosopher Ernst Bloch. *Heimat* for him is also the vanishing point of a political process in which history is quasi the premonition of *Heimat* as something yet to be created – the 'Not-Yet' that shines into all our childhoods (Bloch 1978: 1628). The historicity of *Heimat* is thus not a backward looking one, as the debate on imagined communities and invented traditions often suggests, but rather forward in its orientation.

As a utopian concept, *Heimat* is political-teleological and action oriented in an entirely progressive sense, regardless of whether we are dealing with the utopian socialism of Ernst Bloch or with the 'backward-facing progressiveness' that Harm Klueting (1991: vii) has identified in parts of the German *Heimat* movement of the nineteenth century, whose references to the past were not so much serving the romantic transfiguration of the same but rather constituted an attempt at rationally creating conditions of being (*einen Versuch rationaler Daseinsgestaltung*) that offered an alternative to industrialisation. It should be noted, however, that even attempts to frame a dynamic-dialectical conceptualisation of *Heimat*, which rectifies the distortions and contradictions that have emerged in the process of civilisation, have so far tended to be caught up in the patterns of conservative or indeed reactionary reasoning (cf. Hartung 1991: 153) – which is why I need to continue to tread carefully as I try to explore the matter further.

Having said that, it should not be overlooked that 'conservative' may ultimately also be progressive, for example when the structures to be preserved open up a historically grown range of opportunities for the future that might otherwise be lost. Kamberger (1981: 29), for example, considers *Heimat* as a condition that one discovers, but which one must then first appropriate through immersion in its history, to be able to preserve it through meaningful designs for the future. The concept of 'progressive' as it has been widely used rests on an implicit postulate of linearity that remains insufficiently analysed. Its normative direction for any historical development may well make sense from the vantage point of our desks but is less convincing when considered from a perspective grounded in local actuality, that is, from the arena where history is made in the everyday mediation of past and future. It is therefore advisable to keep an unprejudiced eye on what academics might otherwise regard as detours, wrong tracks or *Sonderwege* that people take en route to their *Heimat*.

Any research on *Heimat* that acknowledges processes of globalisation, however these may be defined in detail, must necessarily engage with themes such as biography and deterritorialisation, as the social sciences have done for some time (see Morley 2000 for a critique of these approaches; also Willke 2001). Concretely, in my case, such research includes a balancing act between at least two languages and/or cultures and thus becomes itself a relevant subject of research. As a person affected, I am conscious of the obvious significance of these circumstances and try my best to navigate the various cliffs entailed therein. In that sense, this section is also part of a process of self-reflexivity engaged in by a migrant mind.

Heimat and Herrschaft

As indicated in the previous chapter, my own reflections on *Heimat* began in the second half of the 1980s almost incidentally in the course of research on informal economic activities in the west of Ireland. Trying to gain a better understanding of contradictions in everyday economic life and its representations that I observed in the region, I worked on the premise of a determining system of ideology which, following Gramsci, I theorised as 'hegemony' or *Herrschaft* in German.[5] Quite accidentally, searching the library catalogue at Liverpool University, I came across a study of Irish patriotic *Aisling* poetry and related literature, published by a German celtologist (Weisweiler 1943). The time and context of publication initially gave me some cause for concern but this was quickly dispelled as I read the text. Indeed, 'Weisweiler's arguments refuse to embrace Nazi ideology, although the political conditions of his time and his subject might have demanded it' (Jochum 2006: 67).[6] The title of his book, *Heimat und Herrschaft*, gave me the impetus to analyse the tensions and contradictions I had observed in terms of this interplay between *Heimat* and *Herrschaft*. Thus my thinking about economic and wider power relations had brought me to think about *Heimat* as something that stands against the hegemony. But that could only be a starting point.

As a field of social relations, *Heimat* cannot simply be reduced to the biological need for territoriality. Going beyond this territorial fixation of the term, I defined *Heimat* at the time in a dual sense (Kockel 1992: 107). Understood epistemologically, *Heimat* functions as a situationally conditioned microcosmic representation which we use to help us interpret the world we live in – to put it more simply: as an explanatory pattern (among others). In the teleological sense, *Heimat* is 'the historical location of the congruence of contextual actualities towards which the human quest for identity and authenticity aspires'. It should be noted here that this aspiration is not necessarily directed at restoring or preserving any particular past – however virtual that past may be – but may at least equally be directed at shaping an as yet unknown future.

This may be illustrated a bit more concretely with examples drawn from the informal economy in Ireland. Especially in the West of the

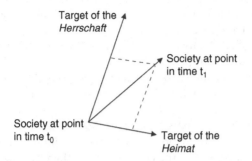

Figure 4.1 *Heimat* and *Herrschaft* as a force field

island towards the end of the twentieth century, the socio-economic foundations for the emergence of new identity regions have been created through community co-operatives, immigrant 'drop-outs' and various relic structures such as *gombeen*[7] (patron–client) relationships (Kockel 2002a). Using statistical data on economic and social development, change can be analysed at county level. This can be done with the aid of a simple vector diagram as we know it from physics, as parallelogram of forces, and as Wiegelmann and others have used it in European Ethnology (Figure 4.1).

The development of a society over a period t_0–t_1 can be measured according to specific indicators that can be consolidated into a summary vector. Thus development paths may be traced and graphically represented for individual regions. As a next step, the key targets of government policy can be plotted using the same set of indicators, and then translated into a *Herrschafts* vector that proceeds from the *status quo ante* and points to a state of affairs that the society ought to have reached over the period, according to the hegemony. While it is not possible to deduce a straightforward *Heimat* vector from this – at least not without employing the *ceteris-paribus* clause beloved by economists – it is nevertheless instructive to compare regional systems with one another and with the national system (Kockel 1989b). For the 1970s – the decade preceding the field research that led to these considerations – it is noticeable that the direction and extent of socio-economic change measured by this model for the West of Ireland deviates drastically from the overall picture for the Irish state. The values for County Galway are particularly interesting. Here the labour market indicator, with an expected value between 0 and 1, shows a deviation of –31, indicating that the regional labour market not only does not comply with the standard economic model – a value close to 0, which would not be unusual – but also that factors such as migration or the level of participation in the work force are determined altogether differently.

A more detailed discussion of the model would lead too far away from the theme of the present book, but it is worth noting that the Galway region flourished economically during the 1980s and, in stark contrast to the rest of rural Ireland, recorded an increase in population. Some other counties also showed negative indicator values but none was anywhere nearly as extreme as the value for Galway. Global factors that affected the entire island may well have facilitated their 'deviant' development paths, but in the case of Galway there seem to have been significant local factors at play – which an appropriately calibrated *Heimat* vector ought to be able to capture.

During and after my field research in the West of Ireland, I studied the German Youth Movement, because from this movement and its successor movements there were and still are numerous connections to the alternative culture in the region (see Kockel 1991, 1995). In the course of these studies, I made a note (and filed it for a while) of Reulecke's (1991: 4–5) observation that the *Wandervogel* had set out to discover the immediate and wider environment of its members – that is, *Heimat* in the conventional sense of the word – as *abenteuerliche Fremde*, which could be translated as 'adventurous foreign territory', and that experiencing this 'territory' was linked to the educational ideal of teaching oneself (op. cit.: 8). In a different context, I later picked up that note and followed its lead a bit further.

Heimat and *Fremde*

The main focus of my empirical research in the 1990s and early 2000s was in the field of culture contact and conflict, especially in border regions and among migrants in Europe under the aspect of European integration. In earlier work I had rejected the term *Fremde* as an analytically useful antithesis to *Heimat* – as mere 'Non-*Heimat*' it did not strike me as a particularly fertile concept, and the idea of juxtaposing *Heimat* as resistance to *Herrschaft* was clearly more useful. However, I found myself increasingly thinking about *Fremde* as a possible term for 'the other'. But such exercises in translation are fraught with conceptual traps. While it may well be an analytically more powerful term than the commonly used German equivalent, *der/die/das Andere*, which simply indicates difference, a 'non-self', anyone translating *Fremde* as 'the other' without regard to the definite article loses opportunities for subtle terminological differentiation (see Kockel 1999a), which is at best blurred in a language that uses the same form for all grammatical genders.

Even without exploring this differentiation in detail, it may be noted that *Heimat* can indeed be conceptualised as complement to a *Fremde* that appears quite different in nature. This is because, with increasing mobility in the course of societal modernisation, the relevance of *Heimat* as something left behind, and to be regained in the anonymity of the *Fremde*, has grown (cf. Klüter 1986). *Fremde* is used here in its feminine form, and within a spatial frame of reference, to designate a location rather than a female person. When *Heimat* serves to locate identity and thus determines a space defined by boundaries, then *Fremde* can be understood in contrast as open, indeterminate space. In European ethnology, the concepts of a 'closed' and an 'open' horizon have been in use for some time. The indeterminate, open space in English is called 'frontier', a term related to 'in front of', that which lies before us – between the perceiving subject and the horizon. Thus *Fremde* can be understood as that which is outside of our selves but within our respective, specific version of the world. In this way, the *Fremde* is a part of, and indeed essential for, the self-determination of each individual or group. This means that the distancing of 'the other' – what is nowadays usually referred to as 'othering' – can be a strategy of translating the unknown into a familiar *Fremde*, which brings 'the other' closer but not into the *Heimat* – leaving it to be *unheimlich* (uncanny; lit.: unlike home).

Regional identities are normally determined territorially; however, in an ever-growing number of cases, that aspect is becoming blurred or comparatively unimportant. Identity is increasingly defined not via personal roots in a particular, relatively fixed place but via labile stations of experience along a life path. Stuart Hall already in the 1990s proposed to shift the focus 'from "roots" to "routes" as a way of thinking about culture' (Hall 1995) and argued that this change can be observed not only in the course of decolonisation but also in the entire diaspora experience characteristic of late modernity. In the regions of Europe we can observe these tendencies, which cannot always be explained with reference to the model of an 'internal colonialism' (Hechter 1975). In the case of Basque identity, where the territorial embedding has been superseded at least rhetorically by the concept of a Basque working class, developed since the 1960s by the militant nationalists, that model may still be applicable (Kockel 1999a). But the Neo-Danish movement in Schleswig, for example, does not fit the model. After the end of the Second World War, the Danish minority in Südschleswig grew to such an extent that this could not be explained by migration and natural increase. The incontestability of ethnic self-ascription in the borderland has long since been legally enshrined. *Gesinnungsdänen* (Danish-minded), a term coined in the late 1940s, has become common as a descriptor for those inhabitants of the region who – for whatever reasons and regardless of ethnic descent – want to be Danes. South of the border, in German Schleswig-Holstein, the so-called *Reichsdänen* – that is, visitors or immigrants from Denmark – are often regarded more like foreigners by the members of this Danish minority (Kockel 1999a). On the one hand, the territorial references for identity in this region are becoming increasingly vague; on the other hand, it appears that a form of regional identity has arisen here that transcends the nationalisms of the nineteenth and twentieth centuries, located in the fluctuating space of the *Mischgrenze*. While in Schleswig that process has been relatively free from conflict, elsewhere the situation has been rather different. In Northern Ireland, for example, the politicisation of the cultural landscape as a symbol of ethnic differences, as discussed in Chapter 2, is accompanied by a kind of renewed mythicisation of history (Kockel 2001a) that might encourage the intensification of conflict in the region. While the current peace process has contained any stirrings in that regard, the signs for the future are ominous – in November 2009, the Independent Monitoring Commission (www.independentmonitoringcommission.org), set up in 2004 to 'to help promote the establishment of stable and inclusive devolved government in a peaceful Northern Ireland', reported the highest level of dissident republican activity for nearly six years.

The re-mythicisation of history in Northern Ireland demonstrated particularly clearly that in addition to the spatial dimension, which theoretical analyses have been primarily focused on in the past, *Heimat* also has a temporal dimension of at least equal importance. It is therefore necessary to conceptualise *Heimat* as a creative process of orientation in time (cf. Tschernokoshewa 1998). The ideal of a postmodern-multicultural society promotes the dehistoricisation of the foundations for identity and thus accelerates the loss of tradition by rooting identities in the 'now' without any sensitivity for the necessity of deriving them from any past other than perhaps one's personal biography. This ideal obscures a dual discourse full of tautologies. On the one hand, 'good' traditions are regarded as inauthentic because they were invented, which means that 'authentic' traditions can only be bad and are therefore dispensable and must be disposed of. Implicit in this assessment is the norm that identity must not build on the past, or should at least avoid it as far as possible. All that may be permitted are representations of the past, mimesis, which can be declared as a construct – and hence as inauthentic. That we are dealing here with a political value judgement that underpins many studies and anticipates – if not to say prefigures – their results is all too frequently concealed. As an alternative to hegemonic culture, non-territorial *Heimat* can be cultivated in situations of culture contact, that is, in the *Fremde* (frontier). The 'drop-outs' and *Wandervögel* in the West of Ireland, mentioned earlier, are a case in point, as are those 'young Europeans' without any great local ties, who especially since the late 1970s have been moving in the opposite direction, from Ireland to the 'European mainland'. Defining their identity via events and relationships that have been important in their individual life history and locating it in the unbounded space of their mobility, these migrants contribute not only to a redrawing of the boundaries of European regions but also especially to a redefinition – in the fullest sense of that word – of the concept of region itself (Kockel 1999a). The ethnic *Fremde* is, in this sense, a necessary condition for the coming home of contemporary European societies, not to any fixed territories of the past but towards an appropriation of themselves and their social historical 'biographies' into the future. Terminologically, the term *region* is less locally grounded than the term *nation*, denoting more the reach of a particular hegemony (*regio*), and is therefore related to the concept of *Fremde* as frontier with its fuzzy boundaries. By contrast, *natio* literally designates the birthplace, and thus is more clearly definable in spatial terms. *Heimat* as *regio* has a certain resemblance to what Hermann Bausinger once described for the nineteenth century as *freischwebende* (free-floating) *Heimat* – a locality detached from concrete territorial references. 'Region' should be understood here not as having a fixed substance but as social space, that is, as a space that has been produced historically, is continuously built anew and remains changeable. It becomes relevant for identity as an expression of the collective networks of experiences, interests and communication (cf. Tschernokoshewa 1998: 166).

Contrary '*Heimat*' and regional development

Since the mid-1980s, more formally after the Maastricht Treaty from 1992 onwards, a change of direction in EU regional policy has brought culture and identity as resources for the development of a 'Europe of the regions' into view. As a consequence, *Heimat* in that classic sense of territorial belonging has once again become an object of politics; that does not necessarily mean it has degenerated into a toy for hegemonic interests. Reference to *Heimat* as an 'object' (lit.: thrown against) to national and European politics can maintain its subversive element; it can, furthermore, demonstrate alternatives to conventional ideologies and praxis. Informal structures and practices, as I discussed them earlier for the West of Ireland, can be observed elsewhere, too, for example the Local Exchange Trading Schemes (LETS) in England (Kockel 2002a). By strengthening local cultural milieus, these schemes contribute to the development of local society and thus gain relevance for identity. In that regard the often voiced criticism, that LETS are not viable as an alternative to the monetarised commodity economy for society as a whole, is missing their key point. Within their respective locally comprehensible space, such structures serve the successful provision of goods and services and thus help secure individual provisions or even a certain prosperity where structures and processes of the monetarised commodity economy fail. This form of informal economy has therefore been referred to in German as *Notbehelfswirtschaft* (makeshift economy), and while its local merits are well recognised, Bernd Warneken (in Jeggle et al. 1986: 18) has rightly warned against an overly enthusiastic interpretation of such practices. However, it may be noted without romanticising the circumstances that makeshift economy can have effects on the social and economic structure of individual regions that are only insufficiently captured by conventional methods of social and economic data collection.

This includes, for example, the shift in the demographic structure of County Galway, which I referred to earlier, where the non-emigration of 'superfluous labour' and the simultaneous immigration of people who were not integrated into the official labour market ran radically counter to market signals. The region's decidedly cultural 'otherness' turned out to be an economic factor that is difficult to capture with conventional analytical tools. Other European regions, too, are cultivating their history of contrariness as a marker of identity and a resource for development. In the south of France, for example, the region of Pays Cathare has defined itself since the 1980s in a dual sense via its contrariness – on the one hand with reference to the Cathars, a heretical community in the Middle Ages that had its geographical centre in the region, and on the other hand through its broader cultural rootedness in the Languedoc with its self-image that is defined in contrast to 'Paris'.

To cheerful observers of globalisation that may all sound very nice since this kind of contrariness chimes well with the ideal of glocalisation, the local appropriation of global developments. The second modernity may indeed promise us all a glocalised homecoming. But I have my doubts. Even in a contrarily glocalised *Heimat* it seems to be the case that the dirty jobs are done by others than those who like to talk so much about glocalisation, and that these 'others' are increasingly those made homeless by global modernisation. Where *Heimat* appears as everyday resistance, this may well be interpreted positively, as an aspect of individual or group-specific liberation – even where this refers to relations with the *Fremde*. But such an interpretation is by no means inevitable. The juxtaposition of *Heimat* and *Herrschaft* is often celebrated precisely by those who reject the juxtaposition of *Heimat* and *Fremde*. The reasons for this are persuasive. And yet reading the contrast of *Heimat-Fremde* exclusively in terms of xenophobia is far too narrow an interpretation. It seems to me rather that this analytical short-circuitry is part of the postmodern project of destroying identity by discourse. *Heimat* can indeed be ideologically abused, precisely because it is not just about nostalgia but also – and significantly – about the historicity of designs for the future. That is why nowadays, more than ever before, we need to engage critically with its cultural realisations in the everyday.

Years ago, Dieter Kramer (1997) spoke of the necessity of cultural studies, and the thinking back to the roots of European ethnology in political science, which I detect in the work of Reinhard Johler (1999a, b) and others, takes up Kramer's lead. European ethnologists can show that in everyday life homogeneous cultural regions are a fiction. Konrad Köstlin has remarked that regional culture appears attractive due to its ability to suggest timelessness precisely because of its historical references. European ethnology can offer a constructive critique of such historical narratives, just as it can counteract the de-historicising of the foundations of identity in a postmodern-multicultural society rooted only in the 'now'. We should be reluctant to declare the quest for *Heimat* in a Europe of free-floating regions as an expression of xenophobia without examining it more thoroughly. Political correctness may demand that differences are not explained, but are instead explained away. We should not play that game. Any reference to cultural differences can and will be politically abused, as so often before. But that has less to do with those differences than with a lack of political culture. When we talk about differences, whether under the 'unity in diversity' slogan of the EU or in other contexts, then we also must take up the debate on ethics and responsibility once more. How we research and write about *Heimat* can influence the ways in which people experience *Heimat* today – and tomorrow (Maase 1998: 69–70). European ethnology can thus play a critical role in turning the world into *Heimat*.

With the fall of the Iron Curtain, the everyday world has become bigger for people on both sides of that dividing line who have since been engaging in new, if often somewhat uneven, forms of dialogue in their attempts to re-create a *Heimat* Europe of sorts. While much attention has been given to the legacies of decades of Communist totalitarianism, far less has been said about the totalitarianism of the Market – or, to use Hardt and Negri's (2000) term: Empire – on the western side of the Iron Curtain. There also seems to be very little reflection on the occasionally noted observation that East and West cannot simply go back to the Europe that was before the Iron Curtain, because the world context has changed dramatically since then, and the East and West that were divided in 1945 were not the same as the East and West that since the 1990s have again been able to interact more extensively at the level of the everyday. Much of this interaction, not only in the early years (Burgess 1997), has been in the rather one-sided form often referred to as neocolonialism, whereby the West has imposed its models and ideals on the East. Arguably, many in the East have been only too eager to embrace these models and ideals, often uncritically. But there have also long been calls for a more balanced dialogue.

Towards a dialogue of dissent

Developing the dialogue between East and West in Europe is an important undertaking. The Institute of Irish Studies in Liverpool, where I worked from 1988 until 1999, was set up as part of a different East–West dialogue, one where cultural differences are perhaps as great, but people assume that at least they speak the same language – which, in fact, they do not, despite superficial appearances (Dunlop 2007).

William Butler Yeats once said that 'Ireland belonged to Asia until the battle of the Boyne'[8] in 1690. This image points to the perception of Ireland as the last remnant of the great 'Western frontier', which now has been replaced by an eastern one. Some European ethnologists detect a resurgence, in the post-1989 period, of old boundaries: 'After the "end of systems", there are conflicts mostly along the old fault lines between the three major European culture areas' (Roth 1996: 8). In rather broad characterisations, these are often identified in both religious and ethnic terms, as Orthodox-Slavonic, Catholic-Latin/Romanic and Protestant-Germanic – although there are clearly more exceptions than would be necessary to prove the rule. To these three must be added at least a fourth, the Islamic area, Osmanic in the East and Arabic in the South and West (Geiss 1993).

Here religion is indeed significant as an ethnic category and marker, quite regardless of any actual religious practice. Journalists reporting on the war in Bosnia frequently used the distinction between 'Serbs, Croats, and Muslims'. This terminology has implications. More consistently, the parties in conflict might have been described as 'Orthodox, Catholic and Muslim', a practice regularly applied in an unreflected manner, when the labels 'Catholic' and 'Protestant' are used with reference to the Northern Ireland conflict. Naturally, this would have suggested an essentially religious background to the conflict. Religious conflict, at the end of the twentieth century, was widely considered to be an anachronism, something belonging to a pre-modern, pre-Enlightenment world, or to the global periphery. And yet when forced to identify the three main parties to the conflict in Bosnia, journalists resorted readily – as they have long done with regard to Northern Ireland – to the use of a religious label where an alternative ethnic definition, other than the straightforward one of 'Bosnian', was not available. Of course, the other two parties were being referred to as 'Bosnian Serbs' and 'Bosnian Croats', respectively, and to speak of 'Bosnian Bosnians' would not only have been a little awkward but might even have been regarded as making a political point concerning the legitimacy of territorial claims that would have been unacceptable on the diplomatic stage at the time. This raises crucial questions about ethnic classification.

The historian Imanuel Geiss (1993) divides Europe along a North–South and an East–West axis, into a Roman South and a non-Roman North, a Latin West and an orthodox East. At the intersection of the two axes, and stretching towards the East, he locates Central Europe as an 'interface territory', the western boundary of which he dates to the period of Charlemagne. This boundary coincided largely with the line marked in the last century by the Iron Curtain.

The Iron Curtain, while it lasted, was – at least superficially – an integrating force of clear definitions and certainties. After it fell, once ostensibly homogeneous nations began to disintegrate into ethnic frontiers. In the East, this found expression in the resurgence of nationalistic movements, while in the West, we witnessed an acceleration of xenophobia and racist tendencies. In the post-Communist era, nationalism in the East often seems to be used to obscure the qualitative distinction between totalitarianism and democracy, between the old socio-political structures and the new ones which are supposed to have taken their place (cf. Gabanyi 1992).

National cultures – do they actually exist in the sense the term is usually understood? What do we mean by 'nation'? Earlier in this chapter, I have considered these questions with reference to Germany. If linked to the territorial state, as in the nineteenth-century concept which has come to dominate most nationalist ideologies, then national culture is necessarily reduced to a small number of sharply defined symbols (Niedermüller 1994). Such 'national' cultures are becoming increasingly threatened as their maintenance is difficult in a free market system of large-scale migration. In some cases, they have also been thrown into disarray by forced migration, for example in the Baltic states during the totalitarian period, which is now generating potential for ethnic conflict. In the West, there are similar cases. Some Basque nationalists, for example, would argue that the large-scale relocation of Spanish workers to the Basque country under Franco tipped the ethnic balance significantly in favour of non-Basques. The radical nationalists responded to this situation in the 1960s by redefining Basque identity, using the more inclusive concept of *Pueblo Trabajador Vasco* (Basque working people) rather than place of birth. The Ulster situation, on the other hand, is a salient reminder of how exclusive ethnic ascription can perpetuate conflict over a very long period of time. Here the loyalist fringe parties in particular have, in recent years, attempted a similar redefinition of identity (see, for example, Kockel 1999a).

As the corporatist ideology of the Plan has been replaced by the individualist ideology of the Market, and people in the East have become more attuned to the latter, a political system which for many in their everyday lives was perhaps inconvenient but not unbearable has become, in retrospect, quite intolerable. Ethnological and social psychological research suggests that individualistically minded people – to put it simply – just want to be left alone, to get on with their lives and, as long as they can make a decent living, care precious little about the system of government. This is where the 'wall in the heads' comes in – the often heard assertion by East Germans, for example, that they wanted to improve socialism, not to abolish it. The old system, they admit, had its faults but, they argue, it was not altogether wrong.

The 'end of history' (Fukuyama 1992) is supposed to have been brought about by the victory of market liberalism over totalitarian ideologies. But we need to define exactly what we mean by 'totalitarianism'. Is not an economic totalitarianism potentially even more dangerous to culture than any political form? Economic totalitarianism has the capacity for diffusing the potential for dissent and resistance, whereas political totalitarianism crystallises them and, thereby, creates its own antidote. Soviet-style totalitarianism made much of historical and dialectical materialism, but the real materialism is found in the presumed rationality of the free market economy.

Free market rationality is based on the assumption of the rationally acting individual. This does not necessarily involve financial calculus.

Imagine sitting in a line of traffic when the driver in front of you starts to reverse to let another car out of a parking spot. You tip your horn briefly to alert him that he is coming close to your own vehicle. After the manoeuvre, the other driver jumps out of his car and walks up to you, evidently irate. What did you blow your horn for, he wants to know. You explain quietly that you wanted to alert him as you thought he might not have seen your car. To which he quips that his vehicle is fitted with electronic sensors and walks off in a huff. Two people acting perfectly rationally nearly ended up in a fight. Why? Because you could not know what equipment he has in his car – and he made no allowance for that. While this is not, strictly speaking, an economic example, there is a hegemonic school of thought that claims economic rationality underpins all everyday interactions. If that were indeed the case then the 'enlightened self-interest' presumably underpinning economic rationality arguably ought to include consideration for others. But would rational *homo oeconomicus* only show consideration for others according to strictly economic, that is, financial rationalisations?

Now imagine you are pulling into a petrol station with three pumps. One side is busy, on the other side there is a minibus in front of you. The driver stops in the middle of the pump island, blocking access to all three pumps, and proceeds to the shop. After a considerable time he returns with bags of groceries and you indicate that you might like to fill up your tank. He comes over to you and asks what your problem is. You reply calmly that there is ample parking space away from the pumps for customers who are not looking for petrol. He barks: 'I pay my road tax like everyone else, and I can park where I want!' Then he returns to his minibus and after a further delay drives on. His response is quite rational, in a way, following the 'user pays' principle – he has paid to use the road, now he is free to do as he pleases. Do his rights as a taxpayer absolve him from showing consideration for others?

Following the Harvard philosopher John Rawls, many economic theorists adopted 'rights' as the basic principle of economic behaviour after critics demonstrated that 'profit' was not a universally sufficient explanation for human behaviour, while 'utility', its successor as the basic principle of economic behaviour, can be used to explain just about anything, and therefore explains nothing. It seems almost ironic that economic theory has no coherent concept to explain economic rationality. Unfortunately, the core assumptions behind *homo oeconomicus* cannot be easily dispensed with; as Holy and Stuchlik (1983: 116) have noted, the analysis is only able to proceed if it presupposes 'the purposiveness, intentionality or goal-orientation of human behaviour'.

However, if the analyst filled these assumed qualities of behaviour with specific content a priori, then 'irrationality' might be nothing but non-compliance of the actors with the expectations of the researcher. Against such intellectual arrogance, Holy and Stuchlik caution that rationality is always bounded and determined by the context of any action as known to the actor.

Dissenting economists have long argued that, while there is a degree of superficial choice in market economies, consumers are hardly ever confronted with choices involving the wider repercussions of their behaviour. A famous example is the small town bookshop (Hirsch 1977). As discount shops offer cheaper books, the consumers will make the 'rational' choice of buying their books where they cost less, bringing about the inevitable closure of the bookshop, but they are never given the choice of keeping the bookshop, with the range of services it offers, in town.

At the time of writing, the Northern Ireland minister for the environment, Edwin Poots, turned down a planning application by retail giant Tesco to build its largest store in Ireland near Banbridge (*Belfast Telegraph*, 1 December 2009). Local traders welcomed the decision. With a share of about one-third of the groceries market, Tesco enjoys a virtual monopoly position. Part of the basis for this dominance is that the company has been able to expand incrementally, by buying individual shops and smaller convenience store chains. These individual purchases are usually assessed as unproblematic by the Office of Fair Trading (OFT), despite that organisation's general view that a share of more than eight per cent of any market can lead to 'anti-competitive distortions ... along the supply chain' (Boyle and Simms 2009: 116). For example, with regard to the acquisition of six former Somerfield stores from the Co-operative Group Ltd., the OFT concluded that it had 'not identified any competition concerns at a national level, given the negligible accretion to Tesco's market share of grocery retailing and to its position relative to its suppliers'.[9] It does not require a background in mathematics to realise that a small addition to an already large base is not going to make much difference to that base in percentage terms. Indeed, the more of these small additions are acquired, the less significant each addition becomes as the base grows slowly but steadily. The problem is well recognised. Already in March 2004, Friends of the Earth called for a moratorium on further expansion of this kind. The rejection of the proposal for Banbridge is the latest in a string of setbacks that Tesco's expansion plans have suffered recently (www.tescopoly.org).

What does all this have to do with Europe, and with an ethnological perspective to it? European integration is decisively driven by economic motives and a particular western model of economy is being exported to the former Soviet bloc where Tesco, for example, has long arrived in some strength: the largest Tesco store I ever visited was not located in the British Isles but in Kraków. Glasman (1996) examines the 'unnecessary suffering' caused when Poland, after 1989, chose 'market utopia' rather than the historical alternatives available to it. The transformation of the market place, and the ideology that underpins it, deeply affects the way we live our everyday lives, what we eat, how we shop – what and how we consume. Thus it fundamentally alters our cultural–ecological relationships. This rationality of the free market economy is supposedly built on common sense considerations and fundamental existential interests – for example, the need for 'jobs' in order to provide food, shelter and lifestyle – under the guise of free choice. If you want all these things, then you must acknowledge the ideology as true, that is, you must pretend that the normative is empirically validated. Therefore you have, in fact, no choice but to accept the ideology with all it entails. Free market economics, like Communism, is, deep down, nothing but a fundamentalist religion.

Creativity, far from being the natural product of the free play of market forces, is always dissident by its very nature. To be creative, therefore, means to disagree with the status quo in some way, whatever the political system. If this were not the case, then there could be no innovation because innovation is, by definition, an expression of difference, a change from what was there before. In the ideology of the free market economy, innovation is reduced to a purely instrumental concept, as technological changes which facilitate the generation of financial profits on a larger scale, over a shorter time span, or both. All other innovations are dissident, potentially damaging to the system of belief and must be suppressed. Western-style market economics as a form of oppression is, of course, rather subtler than the command economics of Communism. It is far more confident in its efficiency at delivering its material promises and consequently does not require forms of overt, political oppression to make people conform.

And yet, where the ideology does fail to deliver, people also succeed in creating niches of dissent, pockets of resistance within a hostile environment, as my work on the informal economy (Kockel 2002a) and many studies in other western societies demonstrate. Western analysts invariably seem to have interpreted observations of informal economy in Eastern Europe as reassurance that the Soviet system inevitably creates some form of capitalistic resistance from within. Participation in such activities was seen as nothing more than an attempt to maintain a certain standard of living under a regime which failed to deliver its material promises. But this could equally be said of Western countries. In the East, however, informal economy seems to have been an integral part of the policy design. Communist states apparently used it to stabilise their total economy and the initial waves of privatisation in Hungary, for example, rather than indicating a fundamental change of ideology, reflected primarily the capitulation of the bureaucracy in the face of mounting administrative difficulties in adequately controlling informal economy for policy purposes. If the magazine *The Economist* (20 September 1980), citing the Hungarian newspaper *Nepszabadsag*, reported that at least 70 per cent of all families in Hungary received some part of their income in the informal economy, this was really no indication of any qualitative differences between East and West. Although consideration of former Soviet economies seems to support a correlation between the degree of nationalisation of formal and the extent of informal activities within an economy, reality is not so simple. In Italy, with a rather weak state and relatively low taxation, informal economy is said to be far more widespread than in Sweden, on the opposite end of the spectrum. It seems that informal economy is more a function of other factors that, so far, have escaped conventional analysis.

Criticism of economics is ubiquitous and has a complex history reaching beyond the current 'anti-globalisation' campaigning. Coleman (2004: 17) undertakes to provide 'a history, analysis and appraisal of anti-economics', and he achieves his goals, after a fashion. A quixotic fight against imaginary monsters, but erudite and mostly enjoyable to read, his treatise offers a comprehensive assessment of many major and minor critics of economics. Coleman defines an anti-economist as 'whosoever sees economics as a bane ... not a mender or reformer of learning' (op. cit.: 7) and anti-economics as 'a hostility to only one sort of economics', to what he calls the 'Mainstream' (op. cit.: 8). Not everyone he classifies in this way fits his definition. Some were or are genuine critics, reformers whom he quasi excommunicates as the Catholic Church did with Luther. Throughout the text economics seems coterminous with political economy, but why, then, is the latter also dealt with as a form of 'anti-economics', and why is much of 'dissenting' political economy notably absent: Kropotkin and most of the anarchists, for example, and also the heterodox economics of Kenneth Boulding and others? It may be that, as Coleman claims (op. cit.: 240, n. 20), 'Utopian Economics' is one book that, as such, 'will remain unwritten', but there is a substantial literature, including studies of utopian communities and other local economic systems, that is not acknowledged. Moreover, the anthropological critique of economics is virtually absent, which, given significant advances in that field (e.g. Carrier and Miller 1998; Graeber 2001), seems particularly unfortunate.

Considerable coverage is given to Continental European authors, but 'Third World' authors are largely absent. Like Eric Roll (1978) before him, Coleman interprets the reluctance of the hegemonic Anglophone 'Mainstream' to engage with their ideas as evidence that their ideas were either insignificant – as otherwise they would have been expressed in English – or 'anti-economics'. France is identified as the 'favoured domicile' (Coleman 2004: 88) of anti-economics. Coleman quotes Pierre Mendès-France (French prime minister 1954–5) who, in deference to 'the Anglo-Saxon reader' whose knowledge of economics is so much greater, pointed to the cultural contingency of economics; he does, however, not pick up on the implication that economics may well be a thoroughly Anglo-Saxon cultural pursuit.

Anthropologists and economic historians working with an ethnological perspective have long argued that the economy is culturally contingent. Moreover, economics abounds with superstitions and moral imperatives in disguise, affirming the societal code of values and conduct we are supposed to live by. Thus it serves as a quasi-religious belief system in our secularised world. The legitimacy of such systems is maintained either by their congruence with experienced, everyday actuality, or by powerful interests.

Extremely critical of the overt religious orientation of some 'anti-economists', Coleman fails to reflect on the religious nature of his own attack. He portrays 'anti-economics' as comprising contrary positions: 'It has been criticised for being a rigid system of belief, and it has been criticised for being a squabble of inconsistent opinions' (op. cit.: 11–12). This is, of course, not a contradiction and the most rigid belief systems are often those that consist of a squabble of inconsistent opinions, masking as 'tradition'. The 'Tradition', as Coleman also describes mainstream economics, is defined as 'composed, not of persons, but of ideas', and including 'some who fought on other sides than the current victors' (op. cit.: 9). There is an uncanny resemblance here with church doctrine and the historical positions of, for example, the Franciscans who were integrated into the mainstream while the Cathars ended up on the stake. Coleman's book follows in the footsteps of the Inquisition when the attack targets personal failings – not least in Coleman's depiction of 'anti-economists' as quintessentially mad (op. cit.: 228ff.).

If economics can be seen as a secular religion, then Coleman's text is a powerful treatise against heretics. A full critique of economics as a secular religion – including its heretics – is still outstanding. There is, quite simply, more to the economy than the western propaganda version of a free market economy admits – just as God is so much bigger and better than similarly fundamentalist religions would suggest. In this sense, I would agree with colleagues from Eastern Europe that post-Communism – and I would add: in its craze for free market economics – is a disease, not a stage in the process of convalescence. Dissent and resistance is needed as much now, to face the accelerating 'coca-colonisation' of the life-worlds, as it was under Communism – perhaps even more so.

Between 1945 and 1989, the West and the East, on the global scale, saw in each other the enemy. Now that the West, according to its chief ideologists, has 'won', the East has become the West's exotic 'other', filling a pseudo-cultural void left when the West's own 'Wild West', its own cultural frontier, was won a long time ago by modernisation. Now the East has taken its place, dark, backward, barbaric, yet at the same time awesomely attractive as a *terra incognita* as well as a highly viable location for quick commercial gain. Conversely, the West has become the East's exotic 'other': bright, advanced, civilised. The contrasting mutual images reveal much about those who hold them (Niedermüller 1997): the heterostereotype can be a mirror image of the autostereotype and so the exotic visions East and West have of each other, however wrong they may be in practice, become mutually reinforcing.

Much of the current East–West dialogue in the cultural field is trapped in a discourse of 'othering' – of rendering the familiar strange. This is not inevitable but rather an expression of the postmodern need for constant titillation, excitement, at least in the West. When this fad fades – which it will, sooner or later – when we become less obsessed with 'othering' each other, it is then that we will be able to rediscover the familiarity which we once shared, for example, when the Irish writer Arthur Griffith drew much inspiration from Hungary. Only then will we be able to fully appreciate our differences and see them as a common resource rather than a reason to build new walls. The old walls are still there, in our heads, and pulling them down with gusto, like the physical wall in Berlin, may not be wise. But we can knock a few windows into them, for a start.

Culture and the European futures market

The development of any culture involves the invention of new things and the forgetting of old ones (Bauman 1999: 73). In this context, cultural traditions – culture 'handed down' – are always 'culture in progress' rather than immutably fixed patterns and practices (cf. Kockel 2002a). Creative use of received notions of 'tradition' and 'heritage', encouraging economic growth through cultural means, can generate significant benefits, and regions like Andalusia (Nogués 2002) or Cornwall (Hale 2002) have long tried, with varying success, to reap these. In most cases, an essentially instrumental understanding of 'culture' and 'tradition' underpins such policies, giving rise to the commodification of culture. However, culture as praxis cannot be quite so readily commodified.

In economics, the concept of 'futures' refers to a process whereby profits expected from the exploitation of resources not yet discovered are traded for profit. As culture has become a commodity, is there such a 'futures market' for cultural traditions not yet invented or 'reclaimed', and what role do the anthropological disciplines play in it? Beyond a critique of the retrospective construction of heritage and cultural traditions, can we anticipate narrative 'futures'? In recent years, the EU has been one of the main investors in such 'futures', largely in pursuit of a European identity rooted in a common cultural heritage. Taking the EXPO in Hanover as his starting point, Johler (2002) questions contemporary identity politics and the role of the EU in the production of cultural heritage and traditions, which he describes as 'scarce goods in the field of global economy'. This leads him to postulate 'a more dynamic concept of cultural heritage' and call for 'new thoughts and ideas from us European ethnologists'. In many ways, the signposts for the way forward point back towards the disciplinary history of European ethnology and its as good as forgotten connection with political economy and governance (Kockel 2002a).

For many regions, and the more peripheral ones in particular, cultural tourism has come to be regarded as the solution to problems of economic development. The reasoning is simple. As Western societies grow more affluent, better educated and aware of the dangers of overexposure to sunlight, tourists are seeking out alternative opportunities for recreation, including the contact – mediated or otherwise – with different cultures. Peripheries are rich in culture (whereas centres tend towards civilisation), which they should marshal as a resource for development. 'Culture' in this sense is generally perceived as an endogenously renewable resource by virtue of the fact that 'the locals' have a vital interest in doing everything they can to keep it alive. This interpretation suits the neoliberalist contempt for any kind of subsidy, but it fails to understand that exploiting a region and then leaving it to its own devices to repair the damage is not what the principle of subsidiarity is about. For endogenous development to work, in tourism or any other sector, we need thorough assessments of what each region's resource potential actually is. Potentials may well differ between regions and between various cultural resources within them. We also need to address the question of value if we do not want to end up in a situation where the only cultural traits considered valuable will be those that attract a price tag. Will it be possible to find a way of acknowledging the value of cultural traits that avoids pricing them without falling into the trap of a shallow essentialism?

Using tradition as a force for change is a strategy implicit in much of the contemporary debate on culture and economy at the policy level, not least within the EU institutions. A first glance may suggest that such a strategy serves to emancipate a periphery that has been 'internally colonised', revaluing its cultural heritage, which is thereby accorded status along with other resources such as crude oil or precious metals. However, this new enthusiasm for things cultural, especially where they relate to the past, is far from unproblematic. To begin with, the entire terminology of 'tradition', 'heritage', 'culture' and so on serves to confuse rather than clarify issues (Kockel 2002a).

Cultural goods such as food, clothing and domestic items of all kinds flow rapidly around the world now, thanks mainly to transportation advances and market incentives for global import/export businesses. This is changing the everyday experience of individuals and groups living in some form of modern diaspora. However, the postmodern dictum that 'we all live in the diaspora' signifies an individualistic utopia for privileged Westerners who may indulge in the economic virtue of perfect mobility. It is normative rather than empirical. Moreover, on closer inspection it reveals new identities formed not so much by a fusion that results from genuine culture contact but by an acquisition – both literally and metaphorically – of cultural icons that almost clinically avoids engagement with anything other than the self of the ego.

There is also the issue of what I would call 'cultural entropy'. Georgescu-Roegen (1972) invoked thermodynamics when deconstructing the circular flow model of economics. While one should not take analogies with physical sciences too far, it is illuminating to consider an aspect of thermodynamics and ecology, the entropy law, in the context of culture and the ethnic frontier. According to this law, any system tends towards a state in which all matter is distributed evenly, particles are spread equidistantly and difference and change are reduced to nil. That system state is referred to as entropic – dead, in plain English. In such a system there can be no more endogenous movement, all components of the system have become utterly indifferent. To bring such a system back to life, as it were, an input of exogenous energy is required, and this will create a degree of order, however fluctuating. Promoting difference to a point where everyone is putatively different from everyone else will, in much the same way as the denial of cultural differences that has enjoyed growing popularity in recent years, ultimately lead to a condition of cultural entropy in which nobody is able to recognise familiar structures and meaningful group attachments any more. Cultural entropy thus makes us all into aliens – it is a state of total alienation and entropic individualism. At this point, it is worth returning briefly to earlier terminological reflections. Alienation would normally be translated as *Entfremdung*. This is not the place to explore different potential meanings encoded in the gendering of the term *Fremde*, which can be masculine, feminine and neutral. But it should be pointed out that *Ent-Fremdung* can signify, in terms of the earlier discussion, either the removal of the individual from the frontier or the dissolution of the 'frontierness' of the frontier. In either case it indicates the destruction of any potential for encounter and creativity, and therefore the impossibility of any homecoming, in a world that has become indifferent.

Familiar structures and meaningful group attachments, however controversial and volatile they may be, provide the negentropy necessary for the system to function. Any working system can be sustained with a relatively modest amount of steering, simply by its own dynamic. However, reviving a system that has become entropic demands a purposeful input and this, in turn, entails power and interest. Considering social systems, any system that falls into a state of cultural entropy invariably suffers a loss of autonomy. It is easy, then, for political interests to dominate and mould such a system. The keen advocacy of cultural indifference as an ideal, frequently camouflaged as the freedom of choice offered by the postmodern identity warehouse, may appear disinterested, but it would be naïve of any ethnologist/anthropologists to buy into this delusion. Rather, what we should be asking is, whose interests are being served by it and whether they are as benevolent as their protagonists make them out to be. The importance of bounded territories may be declining but societies in Europe, and elsewhere for that matter, still need organising principles that enable them to function. The trajectories that make up a 'free-floating' *regio* may replace territory as the organising principle of the region, but we should not forget that, so far, that is only a hypothesis. Particularly needed at this juncture are theoretical explorations of the significance of the past for Europe in the twenty-first century, based on case studies of regions where heritage is a key factor in locating identity and a sense of place. The presence of varied modes of transaction in a particular region implies that these occur within a cultural nexus not defined by any specific mode of production (cf. Gudeman 2008). To date, no suitable models for analysing such situations, potentially involving several distinct modes of transaction, seem to exist.

The fundamental paradox we are dealing with in the context of European integration is that nation states in pursuit of integration are becoming key players in a process that actually undermines the nation state. A state is nothing without a populace – a nation state is no 'state' if it cannot call upon a 'nation'. The free mobility of the populace within the state territory is hard enough to police (in the sense of making policy for). Promotion of mobility across territories for the purpose of the market erodes the substance and authority of the very boundaries that define the state. European politicians of various persuasions have responded with the dual visions of a 'Europe of the regions' and a 'United States of Europe'. If globalisation and migration contribute to a loss of regional and group cultural distinctiveness, then the resulting indifference may indeed engender a Europe of sorts, indistinct and motionless. There are undoubtedly those who would wish 'old' Europe all the best (and Godspeed) on that journey. But is this really what is happening? Is Europe, and Europeanness, increasingly characterised by alienation – by us all becoming aliens? Alienation, as I suggested earlier, is the destruction of the frontier as the creative *milieu* and *metier*. Creation implies, if not a creator *qua persona* – we have all heard of the death of the author and still we write on – then at least a creative force. In view of ever-increasing global interconnectedness, it has been suggested that anthropology should turn away from the comparative study of cultures towards a focus on the flow of goods, persons and ideas (e.g. Augé and Colleyn 2006). The argument is persuasive. But where, one might ask, is the human being in all of this? Are anthropologists, is anthropology, losing sight of the *anthropos*, just as ethnologists are supposed to abandon the *ethnos*? To reiterate: whose interests are served if this happens?

5
Fourth Journey – On the Grand Tour: *We Should Remember*

As Europe – especially, but not only its peripheral regions – is being turned into a theme park for globalised tourists, the question frequently arises and is hotly debated, 'on the ground' as much as in academic discourse – is it authentic? The pragmatic response would probably be: who cares, as long as it sells. This makes some sense. Whether Chicken Tikka Masala is an authentic Indian dish or a concoction for the British palate (see Chapter 3) is possibly irrelevant for the owner of an Indian restaurant as long as it keeps the punters happy. But, quite apart from any romantic metaphysics of authenticity, there is a practical, legal aspect to the current vogue for 'authentic' cultural products – be they food items, fine art or literature in whatever language. Under the Trade Descriptions Act, product labelling must not be misleading. It could be argued that anything offered as 'authentic something' must satisfy recognised criteria of authenticity – unless, of course, 'authentic' is not an objective quality but rather, like 'scrumptious', a matter of subjective taste.

For much of its disciplinary history, one of the key objectives of European ethnology was to establish the authenticity (or otherwise) of its objects of study – whether a particular practice, belief or material item was genuinely German, Rhenish, Transylvanian or whatever else was of the utmost importance. Following the incisive critique by the Frankfurt School of critical theory and the *Abschied vom Volksleben* (farewell to folk life; Geiger et al. 1970) proclaimed by the Tübingen School of empirical cultural studies, this concern with authentication was more or less abandoned. It appeared to suit the spirit of the times in the final decades of the twentieth century to turn attention to the alleged inauthenticity of virtually everything cultural, epitomised by the 'invention of tradition' debate (Hobsbawm and Ranger 1983) and the rediscovery of Max Weber's (1972: 237) idea of *geglaubte Gemeinsamkeit* in the guise of Benedict Anderson's (1983) 'imagined communities'. At the same time, away from our desks, libraries and seminar rooms, 'authenticity' became a major social and political issue with economic implications, while social scientists across all disciplines peddled explanations that smacked of the 'false consciousness' accusations issued by earlier generations of historical materialists.

Rather than being an expression of the 'false' or 'distorted' consciousness of people who do not quite know how to live their lives unless instructed by a social theorist, the recent resurgence of popular concern with cultural identity and the authenticity of cultural products can be interpreted in several ways. It may be 'seen as an attempt to redress the historical injustice of state "internal colonialism" felt by many people in marginal areas' (Ray 1998: 16). However, there are other explanations. Modernity was initially perceived by many not as a threat to culture but as a liberating force. Nowadays, few European regions can credibly present themselves as culturally homogeneous units, and in each local population there are some people from whose personal perspective what is called 'indigenous culture' may appear 'alien'. Can the different issues be reconciled in any way? Perhaps the analysis should not ask whether an identity is 'authentic' but to what extent a region or locality has control over its identity and the formation of local/regional economic activity, and to what extent it may be able to deploy its identity for its own needs.

Historicity between heritage and tradition[1]

Attempts by the EU to create a common European identity have attracted much cynicism. Corresponding to the increasing politicisation of culture, there is now in European policy an extraordinary range of initiatives promoting the exploitation of cultural resources as a key to enhancing the social and economic conditions of local areas (Johler 2002), from long-term programmes like 'Culture 2000' or 'City of Culture' to limited-life ones like PACTE (see Kilday 1998) or 'Pleiades' (see Åhlström 1999), and it has been suggested (Thomas 1997: 336) that the 'objectification of culture at national, regional, and local levels', while 'not wholly unprecedented', has 'become singularly powerful'. Markers of local culture include food and crafts, fine art, language, folklore, drama, literature, landscapes and buildings and sites of historical interest (Ray 1998: 3).

Alongside established administrative regions, there is a growing number of cultural regions trying to utilise aspects of their cultural identity for the purpose of developing their socio-economic vitality. There is a certain schizophrenia at play in this process, as regions try to overcome disadvantages, such as peripherality, by integrating into the EU and the wider global economy while, at the same time, looking 'inwards into the cultural system in order to redefine the meaning of development according to values within the local culture' (Ray 1998: 5). To what extent this process contributes to the widely – if somewhat prematurely – proclaimed decline of the nation state is a moot point that cannot be pursued here but it may be noted that the institutional characteristics of the nation state are indeed – if only partially and to varying degrees in different parts of Europe – transferred to both larger and smaller territorial entities than the presently constituted nation states. The vision of 'Europe of the regions' thus represents a metamorphosis of the nation state rather than its outright decline. A key debate in this context revolves around the issue of citizenship, in contrast to nationality and identity, as highlighted earlier in the reflections on the Drumcree dispute in Chapter 2.

In the new rhetoric that became fashionable at about the same time as neoliberalist politicians began to dismantle the welfare state across much of Western Europe, giving a boost to local culture is regarded as providing foundations for social and economic growth. For most – and not only the peripheral – regions across Europe, that has meant promoting local and regional 'culture' and 'heritage' (whatever that may be in each case) as a resource for tourism development in particular. However, the reappraisal of local and regional resources may also revive an ailing primary sector, in particular agriculture and fishing, which can supply raw materials for the production of 'cultural' goods including culinary specialities. Moreover, a growing emphasis on sustainability has meant that the utilisation of regional culture is increasingly expected to enhance rather than diminish a region's cultural resource base. As an element of cultural identity typically perceived as threatened by social and economic development, language may be a case in point here. There is an increasing number of local and regional initiatives, such as *Menter A Busnes* in Wales or *Gaillimh le Gaeilge* in Ireland, that are using language as a resource to generate development which, in turn, enhances the language by spreading its use not just in terms of an increased population of speakers but also to include new areas of application such as IT and the media industries.

In this section, I consider examples of how culture and identity are utilised, under the banner of 'heritage', to promote development. The underlying purpose of such development is, arguably, the fostering of social cohesion in an expanding Europe seeking 'unity in diversity'. In pursuit of this goal, the EU has put its money on what I will call 'public identities', that is, cultural identities projected into the public sphere (as opposed to 'home identities', which are held in the domestic sphere). The EU sees these 'public identities' as holding the potential for inclusion in the widest sense. If one accepts that perceptions and stereotypes are culturally conditioned, it makes a certain sense to initiate policy moves towards inclusion and cohesion by 'working on' people's identities. But identities tend to be more complex and less mechanical than these policies seem to assume, and the gamble may not pay off.

It is not my aim in this section to critique EU policy in relation to heritage and identity, or to attempt a review of the heritage debate of the past two decades or so. Rather more humbly, I am trying to interpret some aspects of the relationship between identity, heritage and tradition in a contemporary European development context. Following a brief categorisation of identities, which provides a framework for the evaluation of the examples that follow, I conclude by considering whether the utilisation of heritage and identity in Europe today leaves any scope for tradition as a progressive force (cf. Kockel 2002a), and what role ethnologists might play in the process.

Home identities and public identities

The anthropologist Marion Demossier (2007: 59) has argued, following Cederman (2001: 10), that 'the essentialist approach to identity formation is driven primarily by background variables' such as heritage, while 'constructivists place more emphasis on politics seen as an active process of identity formation entailing the manipulation of cultural symbols'. From such an 'essentialist' perspective, these background variables produce identities more or less directly, restricting agency to 'articulation of a given cultural heritage' (Cederman 2001: 10). The concept of a European identity, on the other hand, may imply a more constructivist viewpoint: 'Whereas culture relates to forces that actually shape and have shaped Europe, identity points directly to the discursive level where peoples – consciously or unconsciously – create Europes with which to identify', argues Ifversen (2002: 13–14) and suggests (op. cit.: 8) that '[a] culture projected back in time is normally conceptualized as tradition, whereas history is the grand narrative which orders the past'. Heritage can be regarded as an aspect of tradition that has become 'fixated' (Kockel 2002a) – a parameter rather than a variable. For the purpose of historiographing identities, 'heritage' has to be immutable and, ideally, indisputable. In other words, even the constructivist project needs an essentialised ethnic core as its foundation. Perhaps this is why constructivists find it difficult to explain our intuitive essentialisation of culture (cf. Ifversen 2002: 6).

Identity has many facets. For the present purpose I want to distinguish two levels of identity, which I call 'home identities' and 'public identities'. Both are relational (as identities always are, despite what some social theorists may say) but their orientation is different. Whereas home identities are directed 'inward', public identities are directed 'outward' – the former define the individual vis-à-vis himself or herself, the latter project this individual in relation to the outside world. Each level, again, has an 'inward' and an 'outward' aspect. Taking the level of home identities first, these aspects can be described as 'autological' and 'xenological'– conveying knowledge about, respectively, the 'self' and the 'other'. With regard to public identities, I distinguish between a 'performance' aspect and a 'heritage' aspect. Following the discussion in the previous paragraph, the distinction could also be cast in terms of constructivist and essentialist. Since the objectives of EU policy with regard to heritage and identity are inclusion and cohesion, it may be instructive to consider these different identity aspects in that light. This can be represented as a grid (Figure 5.1).

Speaking Gaelic in Northern Ireland may illustrate these four fields: autologically, the performance of Gaelic speech acts affirms one's identity for oneself, while the same act connects with a specific shared heritage. Xenologically, the performance includes an audience who may not speak the language, but understand its significance to the actor, while excluding all – speakers and non-speakers alike – who do not share that specific heritage.

When the EU promotes culture and identity under the heritage banner, it targets the 'identity fields' AH and XP. Identity is performed (constructed) before an audience of 'others' who are an essential part of 'the theatre', whether as accession countries to be brought under the EU umbrella, the friendly rival across the Atlantic or the perceived barbarian threat closer to home. This performance is essentially founded on a common European heritage (whatever that may be) constituted by individuals, groups and regions identifying themselves with certain 'others' who are also, as it were, part of 'the cast'.

From a public policy perspective, treating identities as merely public identities in this way makes sense, and it works up to a point. However, the ultimate arbiter of identities is the individual and the decision over whether or not identity politics works is made at home. And here the emphasis tends to be on the reverse constellation – AP and XH – which

		Public identities	
		Performance identities (P)	Heritage identities (H)
Home identities	Autological identities (A)	AP Exclusive (the acting 'self' excludes any audience – the others)	AH Inclusive (the acting 'self' identifies with certain others)
	Xenological identities (X)	XP Inclusive (an audience of others is needed for the performance of the acting 'self')	XH Exclusive (the acting 'self' does not identify with certain others)

Figure 5.1   Home and public identities

favours exclusion. This need not be confrontational; it is simply an expression of the prevalent spirit of individualism that stresses distinction over sameness and therefore focuses more on what differentiates the individual from other individuals (even if postmodern individuals, on a quest for distinctiveness, tend to become increasingly indifferent in practice).

Before I move on to some examples, a note of caution may be in order. Although 'performance' has been a fashionable concept in ethnology and other social science disciplines for some time, one needs to be careful in applying it in this context. It carries an implication of virtuality: in a performance of *Macbeth*, we do not see the Scottish king and political reformer, but someone who is pretending to be him, playing out a rather propagandistic horror story. If we consider identity as performance, are we thereby attributing it a similar 'as if' quality – identity as something that we do not really have but merely pretend to have? The answer will depend on how we regard historicity, heritage and tradition in this context.

Heritage as a resource: some examples

Attracting tourism has been widely perceived as a remedy for the problems of peripheral regions. The overseas market in particular has been targeted as a growth sector in the development strategies of many such regions. Treating heritage – widely regarded as a key aspect of identity – as a commodity has implications for the cultural framework that forms the backdrop for any development, leading potentially to the alienation of local people from their (supposed) heritage. Musical heritage has been used in many different regional contexts to attract tourism, and thus provides a good starting point for a broader exploration.

Music

A sparsely populated, remote region, Karelia has been slow to open up to the wider world. Like many other cultural regions in Europe, it has been divided by a state boundary imposed by political interests external to the region itself. Finnish North Karelia has benefited from significant regional development initiatives for many decades, and the vitality of Karelian regional identity has remained undiminished. However, exploitation of culture as a resource for the development of tourism has created a situation where the stereotyping of the region's identity for marketing purposes has to be conserved in order to sustain the commercialisation of its culture (Rizzardo 1987: 48).

This stereotyping of regional culture is not a uniquely Karelian problem nor is it limited to the cultural tourism context. In the 1930s, Scandinavian ethnologists developed the concept of 'cultural fixation' to explain how, under certain historical socio-economic circumstances, cultural forms are conserved, with their meaning reduced to merely a stereotypical badge of identity. In popular perception, Karelia has had a long association with traditional music, and the 'Singing House' at Ilomantsi, near the border that divides Karelia into a Finnish province and an autonomous republic of the Russian Federation, is an important site of Finnish national heritage. This provides a reference point for a historical narrative that portrays Karelia as a region disposed towards music and song more generally, thus constructing music as a heritage resource. At the contemporary end of this semi-fabricated continuity, we find many major international events in the cultural calendar being hosted annually by North Karelia, encompassing a wide range of musical genres and including the Joensuu song festival in June and the Lieksa brass weeks in July/August.

Like Karelia, Estonia has a history of conquest and colonisation stretching back for centuries. In 1992, the newly independent Estonian government prioritised tourism as part of a strategy to overcome the country's economic problems. Planners recognised history and culture as key factors in international tourism. Like South Karelia, Estonia had suffered severe repression under Soviet rule, when traditional cultures were paraded at an annual internationalist festival in Moscow while being persecuted at home, in the republics that made up the USSR (see, e.g. Panteļējevs 1991). Tourists from the West have long regarded Estonia as a window to a 'European culture' of the past. History and heritage are acknowledged as major resources for tourist development, but it is observed that '[i]n addition to our rich historical and cultural heritage we also have our everyday activities to carry out' (Ehrlich and Luup 1993: 10). This echoes concerns that in cultural tourism, where local people themselves may be the primary tourist attraction of a place or region, it should be for these people to decide how much of their culture they actually wish to share with the tourists. Moreover, it reflects a strategy for the promotion of cultural tourism that has been pursued with some success in North Karelia, where the repertoire of regional musical events is based not on a celebration of the Karelian cultural past itself, but on a concept of music as tradition characterised by a specific creativity and the originality of its material, or, to turn a famous phrase on its head, a 'tradition of invention' where the emphasis is on the continued development of 'traditional' skills rather than on the conservation of some unadulterated 'heritage'.

An overwhelming majority of tourists visiting Estonia tend to stay in or around the capital city, Tallinn. In the process of opening up other parts of the country, and developing them for tourism, attractions like the Estonian national song festival are playing a significant role. With its origins dating to the 'national awakening' in the nineteenth century, the first song festival was held in 1869. Although Latvia (in 1873) and Lithuania (in 1924) followed Estonia's lead in establishing national song festivals, the link between song and cultural identity has remained particularly strong in Estonia, where the struggle for national independence from the Soviet Union came to be known as 'the singing revolution'. Estonia's recent success in the Eurovision Song Contest is perceived as continuation of this musical tradition and the event in 2002 provided a valuable injection of tourist revenue between two national song festivals. These song festivals offer potential for a skilful blend of traditional culture and modern originality, an approach that appears to be working well in North Karelia, where a strong musical tradition has also given a boost to tourism. The international musical Eisteddfod, held annually at Llangollen in North Wales for more than half a century now, while much smaller in scale, is another example of how a particular tradition associated with the identity of a region, as choir music is with Wales, may be utilised as a resource for attracting thousands of tourists each year.

As in North Karelia and Estonia, the tourist season in Ireland is relatively short by international standards. There has been a significant shift in the nature of Irish tourism, from a North American market primarily in search of 'Irish roots' towards European markets where family links with Ireland are only of minor relevance. Although continuing migration between Ireland and other European countries may in time create a 'roots-seeking' market in Europe, the current shift has obvious implications for the tourism product. The key characteristics of Ireland as a tourist destination are practically the same as for North Karelia and Estonia – scenic landscapes, a quiet and relaxed pace of life, a distinctive heritage and culture and the absence of mass tourism. In the 1970s and 1980s, supported by an international folk music revival, the Republic of Ireland began to utilise its musical heritage as a resource for tourism, extending the season and attracting large numbers of visitors to ecologically vulnerable areas by staging international folk festivals, mainly along its western coast. Music has played a crucial role in developing Ireland as a tourist destination beyond this, not least through the country's repeated success in the Eurovision song contest, which incidentally produced one of the most fascinating examples of 'glocalised' musical heritage – the dance show 'Riverdance' and its various imitations. An extended version of this interlude to a Eurovision song contest went on to tour the globe, becoming Ireland's key cultural export in the 1990s, but it also had a significant impact back home, radically changing the styles of performance in 'traditional' Irish dancing and, consequently, raising issues of authenticity (Wulff 2007).

The importance of musical heritage for regional identity has been asserted since the 1990s in new ways and from an unexpected direction, which, like the popular impact of 'Riverdance', has brought issues of authenticity and legitimacy to the fore. In Northern Ireland, building on the kind of identity discourse discussed earlier in this book, there has been a growing movement demanding the recognition of an 'Ulster-Scots' heritage, involving language, literature and music. This heritage is seen by its protagonists as in contrast to, but not necessarily in conflict with, the Irish-Gaelic heritage perceived as the dominant, if not hegemonic, heritage discourse in the region. At the same time, this newly discovered heritage also is in contrast with the more 'traditional' emphasis on a British heritage linked to the Union with Great Britain as a whole, aligning itself more with a devolved – and perhaps aspiring to be independent – Scotland than with the UK of the past three centuries. The growth of this movement has been augmented to a degree by the 'Good Friday Agreement' of 1998 with its provision for parity of esteem between the different cultures in Northern Ireland (Nic Craith 2001). This has enabled an Ulster-Scots non-material heritage to be performed publicly and with government support – a level of support that, in the face of tight budgets, has been a major bone of contention (Vallely 2004). In 2004, an Ulster-Scots epic musical, intended to match the cultural impact and international success of 'Riverdance' was launched, targeted primarily at the North American heritage tourist market.

Heritage centres

While musical heritage has been an important element in the Irish case, heritage tourism here is linked to broader historical themes presented with a long time horizon. Local people commenting on new heritage projects often express the hope that they would 'bring tourists in', and this is generally looked upon as a good thing, almost as if the tourists' readiness to travel huge distances to a remote corner of the world is regarded as vindicating the region's way of life. Historically, tourism has tended to create mainly lowly paid jobs for local women (Breathnach 1994). During the 1990s, the heritage centre, a postmodern version of the local museum, displaying some aspects of local, regional or even national archaeology, history and culture, was seen as offering better quality jobs with higher pay. From a planner's perspective, heritage centres have several advantages. 'Heritage' is an omnipresent resource, in the sense that anyone anywhere, regardless of social, political or economic position, can claim some kind of cultural heritage. As a postmodern product, heritage is highly flexible and can be readily adapted to changing market requirements. If it then appears that a heritage centre is not viable in the long term, the building usually looks 'better, less depressing, than an empty factory', as a tourism planner I interviewed in the 1990s put it somewhat sarcastically. In the early years of the twenty-first century, many heritage centres – including some like the Ulster History Park or Navan Fort, representing major investment – have had to close down due to a drop in visitor numbers, and although some of these centres have since reopened following further investment, their experience adds more than a grain of salt to this view.

From an ethnological perspective, the economic benefits of such developments are only a part of the wider context. At the applied level, community involvement is a far more important concern, as it indicates the degree to which the version of heritage represented at a particular heritage centre is actually grounded in everyday cultural experience. Heritage centres offer perhaps more dynamic forms of display than orthodox museums but the danger of 'musealisation' – meaning the detachment of material objects and everyday experiences from their real-life context – remains. In the conventional local museum the focus has been on actual cultural objects of the past, whereas in the heritage centre the dynamics of the display, facilitated by modern technology and know-how, take centre stage. When tourists remember the stunning special effects rather than the story line, the success of a heritage centre becomes questionable. Heritage as entertainment does not require any basis in historical facts or a real-life geographical frame of reference. Just like conventional museums, heritage centres by their very nature tend to accelerate the process of cultural fixation.

While sharing certain structural characteristics and being mostly on the periphery of Europe, the cases considered so far have had diverse experiences with the development of heritage. They indicate both challenges and opportunities for EU policy (Nic Craith 2008b). That an expanding EU will find it increasingly difficult to engineer a coherent 'European' heritage identity – perceived as based on a common past – in all but the most abstract terms is a truism that needs no elaboration. However, the Karelian experience suggests that identity may be based on excellence in a field of contemporary international culture, such as music, and need not be based – at least neither exclusively nor strongly – on a glorious past. Present policy in Estonia reminds us that, although history and folk culture do hold significant potential as resources for development, the fixation of some aspects of cultural heritage for purposes outside the sphere of everyday life, for example through EU-funded regional development projects promoting cultural tourism, may ultimately have alienating effects, and the exploitation of heritage only makes sense if it is grounded in the everyday concerns of contemporary people who continue to engage with it. In a Europe that is becoming increasingly polycultural – both through immigration and through the indigenous cultural differentiation celebrated by the EU's 'unity in diversity' rhetoric – Ireland not only demonstrates the need to devise complex narratives of culture and history that remain, in spite of their complexity, widely intelligible across different groups in society, but also illustrates rich examples of how this task may be attempted, thus affording opportunities to analyse why and under what conditions such narratives may work or fail.

Food

Predominantly rural, the French region marketed as Pays Cathare (Land of the Cathars, a medieval heretical movement) has suffered population decline through emigration, bringing with it all the usual problems of a downward spiral in the provision of local services and infrastructure. In the aftermath of May 1968, the region experienced inward migration, both from urban centres in France and from abroad. As in Ireland, immigration contributed to a change in attitude among the local people, many of them no longer regarding emigration as the best option.

In response to the growth in beach tourism, promoted by the French government since the 1970s, activists in the region of Aude began to see the region as a destination for sustainable tourism. Being at the same time a regional identity and a marketing brand, the Pays Cathare label is governed by quality criteria defined for each of the main economic sectors by the relevant professional organisations in conjunction with Aude's *Conseil Général*. The three main sectors are tourism, which includes accommodation and restaurants; professions and providers of services, including mainly artisans and heritage guides; and agriculture and food. The latter is particularly prominent in the new self-image of the region. Traditionally associated with the production of wine, the region is presented as home to a broad palette of distinctive food products. The tourist is invited to sample these at so-called *étape terroir*, places that have been specially designated for the demonstration of how local produce is made. Claiming culinary distinction within a country that itself claims such distinction on a global scale is no mean ambition for a peripheral region with few resources. In doing so, the region has been able to capitalise on the changing preferences of an increasingly affluent society: the rise in tourism, combined with a growing interest in traditional foods and craft products and the quest for encounters with 'authentic' local people and practices. In late-industrial urban society, the local and authentic is once again becoming synonymous with the rural.

The elevation of food as a key ingredient for a new style of comprehensive regional identity has been observed in other European regions too. Jonas Frykman notes that food from his home region of Skåne in Sweden has become 'a more loaded concept than it ever was in the days to which the tradition refers'. The label 'Skåne' is used 'as a seal of quality' for food produced in the region and the area 'has become something far more than a region, it has become a site, a place charged with meaning that does not necessarily have a geographical foundation' (Frykman 1999:16). Here Frykman indicates two characteristics of this identity badge. It is more comprehensive and coherent than the fractured identities favoured by late postmodernity. At the same time, this comprehensiveness and coherence is built on constructed meaning independent of any actual place and its history in a way that appears rather postmodern. Globalisation has brought 'ethnic' foods from around the world into most European regions and, in some cases, rehabilitated domestic cuisines. At the same time, the growth of the organic movement and the farmers' markets indicate an increasing significance of food and its preparation as markers of locally rooted identities. Subtly, the shopping for and cooking of a Sunday dinner is becoming something of a public statement and political act.

Language

The food one eats is a key signifier of one's cultural belonging. For German migrants in the British Isles, for instance, bread has long been at the top of a list of items distinguishing German from British culture. Another important cultural marker is language. Nowadays, we no longer have simply languages but categories of language – official and non-official, national and regional, major and minor, high and low, more widely and less widely used, standardised and non-standardised, less widely taught and used languages, disputed language varieties, ethnolects and so forth (Nic Craith 2000). This multiplicity of terms has reinforced the social implication that speaking a major language empowers the individual. It has given a false validity to the notion of a social hierarchy of languages, which implies that certain languages are not only socially more useful and economically more viable than others but are also inherently superior vehicles for communication. With very few exceptions, European nation states have adopted a single national language, thus marginalising all other languages within their respective territory.

Many of the so-called minority languages are located on the periphery of modern nation states. 'Minority' languages are often a consequence of boundary changes or migration. By speaking such a minority tongue, a group deviates from the norm and could be seen as placing itself on the margins. As the *Euromosaic* report points out, the term 'minority', when used with reference to a language group, refers primarily to power relationships rather than to any specific statistical measure (Nelde et al. 1996). From the 1980s onwards, 'the regional revival throughout Europe has been accompanied by initiatives to bolster minority languages as a component of regional development strategies' (Ray 1998: 12). One reason for this has been that, in their search for markers of cultural distinctiveness, European regions have found that language is 'a powerful means by which one culture can display its difference from all others' (Ray 1998).

Many regions, even those where there has been a positive attitude to linguistic diversity, have treated regional languages as a liability, as something that costs money, rather than as a resource. Policies have been implemented in order to maintain and protect the linguistic heritage. In Ireland, for example, the government created language protection areas called *Gaeltacht*. Protectionist policies like this have often served to reinforce stereotypes of lesser-used languages in terms of a rural–urban, agrarian–industrial, archaic–modern dichotomy, which has led people to disengage from their regional language.

Government support for a language as 'heritage' can be interpreted in different ways. The French government, for a long time a champion of cultural centralisation, has recently recognised regional languages as part of the national heritage and has begun to support some 'minority' languages like Alsatian. The designation of a language as 'heritage' may be a giveaway, however. Culture becomes 'heritage' only when it is no longer current, that is, when it is no longer actively used. In other words, 'heritage' is culture that has dropped out of the process of tradition. The term 'tradition', literally, refers to cultural patterns, practices and objects that are 'handed down' to a later generation, for use according to their purposes, as appropriate to their context. By contrast, 'heritage' refers to cultural patterns, practices and objects that are either no longer handed down in everyday life (and therefore left to the curators) or handed down for a use significantly removed from their historical purpose and appropriate context – such as to attract tourism.

The undeniable cost factor associated with the protection of such 'heritage languages' is frequently resented by majority language speakers. Against this view, it could be argued that language, unlike many other resources, is enhanced rather than diminished by its use. Moreover, as the language environment increases, the capacity to generate economies within that environment is also enhanced. At an individual as well as an organisational level, many entrepreneurs have recognised that where a social and cultural infrastructure has been created that fosters language use and promotes a particular language environment, the economic pay-off has been considerable and has facilitated further strengthening of the language. Many diverse projects have been initiated across Europe, aiming to increase the use of less widely spoken languages in the commercial sector (for examples, see, Nic Craith 1996). In a situation where most of these languages have few or indeed no native speakers, their use becomes very much a performance, deliberate and pointed. However, the same can be said for ethnolects whose status as language is disputed. For example, Ulster-Scots may or may not have thousands of native speakers, depending on your political perspective. If it does, then speaking (and writing) it in a form that emphasises distinction from English is arguably the legitimate practice of a living language; if it does not, then such practice constitutes a performance act that – as performance – differs little from the use of undisputed languages like Gaelic or Welsh.

Authenticity, tradition and mimetic heritage

When we are dealing with culture, identity and heritage, 'on the ground' as much as in academic discourse (e.g. Bendix 1997), the question of authenticity is usually raised. The Frankfurt School initiated a critique of 'the jargon of authenticity' (Adorno 1973) especially with regard to culture, contesting the implication that there is 'something immanent in local culture systems' because the assumption of such immanence would 'deny any agency of human subjectivity' (Ray 1998: 15).

The resurgence of popular interest in cultural identity in general, and in the authenticity of cultural products in particular, may be interpreted in a number of ways. Few regions in today's world can lay claim to being culturally homogeneous; moreover, in any region there are people for whom what others locally may refer to as 'indigenous culture' appears quite alien. The power of definition is where Ray (1998: 16) looks for a solution to the problem of authenticity, which he suggests 'might more usefully be reformulated as a question of legitimacy, i.e., who confers legitimacy on what form of cultural activity?' Ray argues that the analysis should not so much ask whether or not an identity is 'authentic' but to what extent a locality or region has control over its identity and whether 'the cultural identity [can] be tied to the particular territory so as to meet local needs?' Būgienė (2005) locates contemporary Lithuanian identity with reference to what one might call its 'significant others', in particular Sweden and the Soviet Union. Her account of what is effectively the folklorisation of actual historical events makes fascinating reading and indicates how Lithuanian identity, like that of so many others on the political periphery, is defined primarily negatively by reference to what Lithuanian culture is not, rather than by what it is. Postulating a research agenda that follows from his analysis, Ray (1998: 17) sends social scientists, and European ethnologists in particular, back to basics: 'We need to look more closely at the processes whereby territorial/cultural identities are constructed, promoted and protected. ... We need to know more about the specific relationships between place, history and the on-going process of symbolic construction.' More than a decade later, and despite some significant progress, our understanding of those relationships remains vague and patchy.

We are used to thinking about 'tradition' as associated with fixed formations derived from the past (or projected into it) that hold back or corrupt progress. This 'tradition' is invoked by 'yesterday's men' in their attempt to stall innovation and change. If necessary or simply opportune, 'tradition' may even be invented, especially in contexts where anything with an air of antiquity is regarded as venerable by definition. Societies where such mechanisms are strong tend to be referred to as 'traditional' or 'pre-modern'. In this world view, a 'modern' society becomes 'non-traditional' by default. This, then, is the paradigmatic way of looking at tradition and development.

Pertti Anttonen (2005) blows the cobwebs off that world view and shows eloquently where it went wrong. Having established the geopolitical context at three levels, he presents a case study of Finland that suggests interesting parallels with other countries and regions that would be worth investigating further. For example, his discussion of Karelians as 'both Finns and Non-Finns' indicates a strong resemblance between Karelia and Ulster – from a 'nationalist' perspective both are 'ethnic heartlands', yet the actual ethnicity of their inhabitants is far from unambiguous. As European integration and globalisation continue to shape our everyday lives, European ethnology can offer interpretations of 'tradition' that are more differentiated and less static than those used by the nation-building folklorists of yesteryears. Anttonen's Finnish case study illustrates this clearly. Perhaps one day, we will even have a concept of tradition as a force of progress?

Approaches that see the past as a construct of the present have enjoyed great popularity in recent decades. The fashion for the past has given way to the fashion for the present, and all our perspectives need to be adapted accordingly. However, it is only a small step from talking about observable symbolic processes whereby values considered desirable in the present are being projected into the past in order to ordain them with legitimacy, to asserting that the past has had no real values at all, or none that are worth transmitting to the present or – the epistemological consequence of a perspective that sees the past entirely as a projection of the present – that we have no way of knowing what the values of the past were since all we can know are our own projections and interpretations of the same. Much paper has been filled with this persistently fashionable style of self-referential analysis that says much about the authors' capacity to play mental games with themselves but contributes little to our understanding of the world, past or present.

If we acknowledge that 'authenticity' is a matter less of true or false consciousness than of the historical legitimacy of any associated identity claim, we can revisit the 'invention of tradition' debate and, conceiving 'tradition' as process, recognise 'heritage' as a fixation of tradition (Kockel 2002a). This enables us also to recognise that it is not so much tradition that has been invented but rather heritage. Looking at Visby in Sweden, Ronström (2005) dissects the triad of memory, traditions and heritage. His juxtaposition of 'tradition' and 'heritage' as interpretive templates is interesting and worth developing, especially at the point where he poses the intriguing question of whether *kulturarv* might be a conceptual alter ego for late- or post-modernity – as tradition is widely assumed to have been for modernity.

Tradition as a process involving cultural actors always includes the possibility of more or less subtle modification of what is being handed down between generations in order to appropriate it to a changed historical context. Only if it becomes fixated as heritage does tradition cease to imply process and change. From this perspective, the use of the label 'traditional' no longer implies something immutable and eternal but refers to legitimacy derived from everyday historicity. Given the increasing emphasis on cultural resources as a means of socio-economic development, we need to re-evaluate the use of tradition, understood as a creative process of intergenerational cultural appropriation, in regional policy, to avoid its instrumentalisation in stale commodity versions of 'heritage'. This can help to empower localities and regions, and at least obstruct the transfer of established patterns of exploitation from the old industrial to the new post-industrial economy. The alternative would be the postmodern view of identity in the age of globalisation. Following that view, doing anything at all becomes hard identity graft and is no longer self-evidently grounded in the everyday. Identity becomes 'mimesis' (Kockel 1999a: 68) as peer group pressures to conform to the mass-produced individualism of the identity warehouse replace old-style paternalism and imperialism as the forces that colonise our life-worlds.

Globalised heritage and the destruction of traditions

The period of fieldwork reflected in this book happened to coincide more or less with the flowering of postmodernism and its distinctly anti-historical attitude. During this time the original interiors of many historical local pubs were ripped out and replaced by off-the-peg 'heritage pub' designs. A similar fate afflicted many other aspects of both material and non-material culture. Could the postmodern disdain for history (and tradition), with its concomitant 'knock down/build new' attitude, be a reflection of the globally dominant culture since the second half of the twentieth century, that of 'white' US-America, industrially creating surrogate heritage to compensate for its often alleged shortage of history? If Ernest Gellner (1983: 34) is right in arguing that '[t]he monopoly of legitimate education is now more important ... than the monopoly of legitimate violence', then what are the implications of progressive privatisation and commercialisation of education, as envisaged in the proposals for a General Agreement on Trade in Services (GATS), driven by the US world trade agenda and fended off provisionally by the EU? The GATS vision implies a further transfer of power over the social order, from political (state) to economic (business) interests. Cultural traditions are endangered if the way is thus cleared for their commercial 'takeover' by the player who is most powerful economically.

The heritage boom of recent decades may have camouflaged an erosion of European cultural traditions, hiding it behind the smokescreen of 'culture as a resource', a strategy that uses cultural fixation to commodify identity as heritage. Earlier, I proposed to conceptualise 'tradition' as a process that is at its heart about sustainability – about the 'handing on' of knowledge and practices for appropriate future use – and of 'heritage' as objects and practices that have become fixated and have thereby quasi fallen (or have been deliberately taken) out of this process. The distinction is important. If the two are confused, tradition can be represented as static and branded 'bad for progress' – as it has been for some time. This raises the question of whose interest might have been served by the maligning of tradition – especially European traditions.

Most of the examples cited in this section fit with the identity fields promoted by the EU, where individuals define themselves via the constructive performance of essentialised heritages, which form a basis for the regional identity that the performance presents to an audience of 'others'. Both identity fields are selectively inclusive in their own way – autological heritage identities define the 'in-group', as it were, while xenological performance identities need an audience of 'others' before whom they are played out. In some cases, these practices go back a long time. The association of Karelia with musical tradition dates at least to the 1830s, when Elias Lönnrot gathered folklore and turned it into the Finnish national epic about the magical power of song. Likewise in Estonia, the emphasis on music played a key role in the 'national awakening' during the nineteenth century. The same can be said for many other European countries and regions. What makes these two cases different is that in the contemporary utilisation of culture as a resource, music is not fixated as heritage but retains its historicity as a creative tradition. The celebration of 'pristine' folk dances and other heritage is marginal compared to a strong emphasis on creativity and evolving practice. In other words, the regions' cultural traditions are characterised by an enthusiasm for innovation. Although music is seen as an essential root of identity, agency is by no means limited to expressing a *given* heritage (cf. Cederman 2001: 10) but is free to progress tradition.

The same is basically true for more recent examples, such as the Pays Cathare branding that projects 'resistance' and 'unruliness' as essential roots of an autological heritage identity and develops from this a xenological performance identity celebrating 'difference' not just from foreign tourists but also from the rest of France. However, many of the cultural traditions displayed in the Pays Cathare could easily become fixations if they are presented as time-honoured ways of doing things that must not be changed. The historical frame of reference of this 'rebellious' identity is at its core a backward oriented, conservative one whereas the orientation in Karelia and Estonia appears much more forward looking.

The case of Ulster-Scots, which has featured here both in relation to music and in the context of language, is interesting for its 'split' perspective in this context. With regard to both music and language, protagonists have been charged with inventing a tradition. All tradition has been initiated by someone somewhere, and is therefore invented. Thus the charge of invention points to something else – the question of legitimacy. If we accept that all tradition is invented, then legitimacy cannot be derived from any primordially grounded authenticity. As a musician in Galway once said to me, a tradition is authentic if it works. A working tradition, literally, is one that is 'handed on' continuously, both across space and through time. In this regard, the jury is still out on Ulster-Scots.

However, this 'tradition' highlights one critical aspect of public identities more clearly than any of the other cases (to which it applies nonetheless, if to a lesser degree). At the level of home identities, autological performance identity is directed at defining the 'self' to the exclusion of 'others', as is the xenological heritage identity. This is inherent and, in itself, not necessarily problematic. All identity is about affirming what we are and, thereby, what we are not. This is often seen as the crucial issue for multiculturalism and some analysts (e.g. Day 2000) seem to regard a specifically European discourse of diversity as the root of all evil, as if only Europeans knew how to construct 'others'. Conversely, the construction of and emphasis on differences has been proposed as 'the only way to oppose the hegemony of Euro-American thought' (Schiffauer 1996: 62). That other cultures may have other ways of doing this (cf. Kimmerle 2002; also Mall 2000) is an issue that cannot be followed up here. The important issue is the use to which such differentiation is put. In the 1970s, the right to difference was heralded as part of an emancipatory agenda. Emancipation is, not least, an autological performance rejecting heterostereotypes in favour of autostereotypes, thereby denying ascribed inferiority and breaking down hierarchies. When such differences are used to assert essential superiority over 'others', however, the ethos of diversity becomes a major political and ethical problem.

At the same time as culture became a political issue, the study of culture in the social sciences and humanities ironically turned its attention to textual analysis. Contemporary theories of modernity overstate the 'reflexive and subject-oriented nature' of culture and identity and, consequently, cultural analysis 'has found it difficult to incorporate the perspective of action' (Frykman 1999: 22). Focusing on action 'gives room for curiosity about the many reworkings that take place locally, their conditions and constantly occurring transcendences', and after a period when the desk, the library and the Internet seemed to become the primary tools of studying culture, we may now find it appropriate to make 'the local' once again a crucial element in our ethnological fieldwork. Such fieldwork can demonstrate the progressiveness of many cultural traditions and challenge the interests behind the fixation of certain heritages – not least those interests who would like to write off 'old' Europe politically, turning it into a heritage theme park for roots-searching global tourists. Whether or not this fieldwork would affirm any 'unity in diversity' from below is quite another matter. Either way, we should not let an irrational 'fear of difference' (Schiffauer 1996) – however politically correct it might be – determine our research agendas but instead heed Emanuelis Lévinas' postulate of *le droit à la différence*, the right to be different. Along with Martin Buber, Lévinas was one of the leading thinkers developing a dialogic vision of the 'other' in more or less indirect opposition to two phenomena of the twentieth century, mass society and totalitarianism (Kapuściński 2008: 84), which have threatened to destroy places and identities. Rather than denying difference, we need to understand more deeply the ecological and symbolic relationships between place and tradition that provide grounding for cultural and territorial identities.

Re-placing tradition and the folk[2]

At the Lithuanian Folk Museum in Rumšiškės outside Kaunas, we find the usual depictions of rural life, customs and costumes of a bygone age. But this museum has an edge to it. In one of the farmhouses from Dzūkija, you can see an oven that has been ideologically de-sanitised. During the Soviet period it was just an oven – now the hideaway underneath has been restored, one of many that were used until 1904 by book smugglers who risked their lives bringing reading material into Lithuania across the then Russian border from Germany and Poland. If you walk on from this farmstead, you see on a clearing in the woods an unusual-looking earthen yurt, across from a railway boxcar. Inside the car, amidst photos, maps and drawings, countless shreds of paper are suspended from the ceiling, each one telling of a deportee. Between 1941 and 1953, and especially during the first few years after the war, well over half-a-million Lithuanians were deported (Vaišnoras 1991: 50), many sent to the Gulags, some as far away as the Laptev Sea on the Arctic coast of Siberia. Many of them disappeared forever; they were not peasants, just folk that Comrade Stalin and his executors were 'putting in their place'.

Of course, I am using 'folk' here in a less restrictive sense than the meaning it acquired in the course of the nineteenth century. The term 'folk' has been found useful by all kinds of ideologies right across the political spectrum: from the French Revolution to nineteenth-century nationalism, from social democracy and the Labour Movement to Nazi-Germany and the communist regimes, from the *Wandervogel* to the Free Republic of Wendland. 'With all their differences, these ... have all used "folk" ... as a concept to denote an ideologically important community' (Hylland 2001: 18). In practice, that usually meant those who supported the respective ideology were considered part of the folk while those who opposed it were considered *Volksfeinde*, enemies of the folk.

In earlier periods, it had a broader meaning designating ethnic groups and their assumed characteristic differences. Stereotyping lies at the very root of well-rehearsed misgivings about ethnology's role in the past – and rightly so. For much of its history, ethnology has been involved in what Jace Weaver has called 'the gymnastics of authenticity' (Kidwell and Velie 2005: 10). Romantic visions of the folk as 'proud, happy, strong, efficient, god-fearing, singing as they work, perhaps telling a tale as evening falls' are the standard fare (Hylland 2001: 22). Moreover, as industrialisation and urbanisation created a new middle class, this 'freedom-loving, individualistic, and principled peasantry, embodying honesty, honour, and love of traditions ... represented the kind of ancestors the middle class wanted to have in their cultural charter' (Löfgren 1987: 60). Imbued with immutable attributes that made them who and what they were, the identity and authenticity of these peasants were never in question even if a rapidly advancing modernity threatened them with extinction.

Native American academics use phrases like 'an American Indian perspective' to indicate that their people think differently. Even an otherwise very good introduction to Native American Studies claims that this notion 'smacks of essentialism, the concept of categorization that presupposes that people must think and act in certain ways because of their ethnic group' (Kidwell and Velie 2005: 9). The charge of essentialism has also been brought against ethnology and its treatment of the folk. However, this currently fashionable vilification of an 'essentialism' that presumes people must act and think in certain ways due to, for example, their ethnicity is itself 'essentialist' in the very sense it uses the term. How is ethnic group membership defined? Ethnic belonging is circumscribed by the actions, norms and representations (Holy and Stuchlik 1983) shared by a particular group and recognised by members and non-members of that group as identifiers, the significance of which is broadly understood by all within the reference context, that is, by both insiders and outsiders. Someone who does not (wish to) – or for some reason cannot (yet) – share these actions, norms and representations is placing herself or himself on the edge, or indeed outside that group. Where this is not a matter of actual capability, it is an act of choice, or, where initiation rituals are involved, possibly a stage in the process of becoming a group member. Only the critics of 'essentialism' presume the existence of ethnic essences preordained in some 'natural' way. However, labelling certain ideas and avenues of inquiry 'essentialist' is a convenient way of avoiding issues that might challenge fashionable normative frameworks.

If we are going in search of the folk – 'who they are, what they do, what is done to them and what is thought of them' – we will find them 'in a variety of places' (Hylland 2001: 19–20). Different stories can be told about them, including different histories of the concept. These histories intertwine at times, most powerfully so in the context of nineteenth-century nationalism. Nationalists everywhere emphasise the commemoration of their 'own': heroes, victories and, especially, suffering, which is invariably blamed on 'others'. The Lithuanian poet Thomas Venclova (2001: 80) speaks in this context of a *Wettbewerb im Märtyrertum* (competition in martyrdom). In the course of nation building, new rituals are devised to underline the national identity, providing markers of belonging for a population whose taken-for-granted markers are thus being replaced. Peasants become Frenchmen (Weber 1976).

Nineteenth-century nationalist movements postulated a territorial dimension to culture. Places that supported a particular culture were seen as 'composed of people like oneself, people with whom one can *share without explaining'* (Sennett 1996: 179; emphasis added). The folk who populate these places are said to have the strongest cultural identity when they are not aware of having one. Although it 'speaks in the name of ... a folk soul', this notion is actually 'a rule for exercising power' and may equally legitimise revolutionary movements or forms of domination and oppression (Sennett 1996: 180). The Italian nationalist Manzoni described the peasant as having no historical consciousness, as someone who simply is. The folk are silenced by definition, and so the expulsion of the folk from their place in actuality begins with their idolisation by the new nationalists. That idolisation of 'the folk' has been an integral part of the eviction of the folk from their rightful place in history in order to put and keep them in a place (on the periphery, of course) designated for them by the hegemony.

Ethnology and the Enlightenment

But that was not what ethnology was about during the Enlightenment and early Romantic periods. Kant (1784), in his *Reflections of a Universal Citizen of the World*, suggested that feeling at home and deriving stimulation among a wide range of different people was important for the development of the human being (Sennett 1996: 181). For Herder, too, people's differences mattered most because these are what makes us who and what we are (Berlin 1976: xxiii). However, Herder also understood that perceptions of difference may engender ethnocentrism (Sennett 1996: 182). Following his influential *Stimmen der Völker in ihren Liedern* (1773), ethnology developed as an academic subject. Derived from the Greek word for folk, ἔθνος, 'ethnology' – 'knowledge of the folk' – is a genre of anthropological study involving the systematic comparison of the folklore, beliefs and practices of different societies. The public role of ethnology springs from, and is indeed inseparable from, its practice. More about this later.

After Herder, Romanticism connected the prefix *folk* with certain aspects of culture, in particular with *poetry* and *spirit*. By the late eighteenth century the folk were attested both *Volkspoesie* and *Volksgeist*. 'The poetry and the spirit were significant because of three things: they were natural, typical and collective' (Hylland 2001: 19). The ideological elevation of the *Volksgeist*, originally postulated by Herder as a descriptive term, can bring on the beginning of the Apocalypse, as Venclova (2001: 81) observes, but this misuse of a concept is neither necessary nor inevitable. The French Revolution is often seen as the culmination – and in some ways also the perversion – of the Enlightenment. Like its American precursor, it had established the ideal of a nation as a political codex, inspired by a kind of Kantian universalism. The new nationalists of the 1840s rejected that ideal in favour of the concept of a nation created through custom and based upon the beliefs and practices of the *Volk* that constituted this nation (Sennett 1996: 178–9). Contrasting as they are, the two visions actually represent two strands of the Enlightenment. And in a curious way these two strands were held together by many individual scholars, most notably among them the grandfather of European ethnology, Wilhelm Heinrich Riehl.

Ethnology's public role

As noted earlier in the book, European ethnology has at least some of its roots in the Enlightenment subject known as *Allgemeine Statistik*, the purpose of which was defined in 1749 by Gottfried Achenwall as forming conclusions about the wise governance of the state, that is, its application in politics (cited in Hartmann 1988: 12). By the middle of the nineteenth century, popular enlightenment had developed as a philanthropic movement sharing the concepts and methodologies of early scholarship in ethnology, with no clear intellectual boundaries between the two agendas (Hylland 2001: 20). In his influential 1857 lecture on '*Volkskunde als Wissenschaft*' (ethnology as an academic field), Riehl outlined a research programme that made him without doubt a path setter for a new conception of ethnology as an interdisciplinary, integrative approach free from philosophico-cultural blinkers (Zinnecker 1996: 321). His classic study on the Palatines, *Die Pfälzer*, became the template dominating regional ethnographies for over a century and established fieldwork as an ethnological method. In it he also dwelt on the problems for ethnologists that arise, in his view, from the fact that each people, or indeed each fragment of a people, considers itself the centre of humanity. Governing such a people may be rewarding, he said, as long as you abide by its idiosyncrasies, but to describe them is a thankless task, for the more thoroughly one studies their characteristics the more uncomfortable they are with being studied. This is a complaint shared by the enlightened philanthropists who found it 'no easy task to be a friend of the folk' (Hylland 2001: 23) they sought to educate. Even intellectuals with roots in the peasantry could become frustrated with the folk (Hylland 2001: 24). A deep analysis of Riehl's work reveals him, despite his involvement in the 1848 revolution (or perhaps because of it), as anti-democratic, anti-liberal, rather romantic and certainly bourgeois. At the same time, he was groundbreaking with regard to interdisciplinary and comparative research, pioneered fieldwork and a functional analysis of culture, and introduced socio-economic and psychological concerns into the study of everyday life long before all this became fashionable a century later (Zinnecker 1995: 22).

In his book *Die Pfälzer*, Riehl also discussed migration. Some of these Palatines had in the eighteenth century migrated to Ireland via London. Patrick O'Connor's (1989) award-winning history of their settlement, 'People make Places', points to what Richard Sennett explores as a key contradiction in our interpretation of place and belonging, 'between the truth-claims of place and beginnings versus the truths to be discovered in becoming a foreigner' (Heelas 1996: 13–14). The point here is the mutual conditioning of folk and place – they form a kind of corporate identity that is sustained by traditions in their ecological contexts. To say this is neither an invocation of environmental determinism nor a suggestion that the settlers imposed any naturally inherited cultural ways on their new, temporary abode. Yet their relatively brief presence left its mark on the area where they lived in County Limerick (and beyond) just as the area, in some ways, stayed with them and their descendants. The experience of the Palatines challenges the notion of places as purely bound, showing them as connected along 'trajectories of belonging' (see Kockel 1999, 2005). The growing interest in heritage since the 1980s may have rekindled awareness of this legacy – it did not invent it without foundation.

Like many others before him, Christopher Tilley, in an otherwise interesting and stimulating essay, portrays ideas about the uniqueness and singularity of places and landscapes as indicative of people seeking refuge from modernity, 'the purity of ethnic groups and continuity in the face of change' (Tilley 2006: 13). People certainly create romantic images of places and landscapes. I am less convinced they do this for refuge. In most cases that I am aware of, these constructions have economic purposes (see also Frykman and Löfgren 1987) and should be recognised as such at least by analysts who essentialise metaphysical market forces. After some 30 years of reading about the subject, I am growing exceedingly tired of cultural analysts beating the same old drum that used to have 'false consciousness' written on it – any positive evaluation of the past and any analysis emphasising continuity over change is branded as indicative of reactionary politics, emotional regression, or both: an irrational scramble for shelter from the vagaries of the modern world. This diagnosis has become so commonplace and deep-seated that anyone daring to challenge it would find herself or himself immediately relegated to the same politico-cultural sickbed. Under no circumstances must we look for continuities (unless we want to be seen as emotionally retreating into a fantasy Golden Age). Why is the mere postulate of continuity so widely shunned by our cultural analysts? Why are we so obsessed with change as the essential human condition – when essentialism is supposed to be dead?

The notion of change and tradition as fundamental opposites is peculiar to modern Western culture, where 'change alone constitutes history' (Lenclud 2003: 74). Cultures that see repetition, cycles of events, as being more significant than linear narratives about unique events view the past as 'being incessantly reincorporated into the present, and the present as the repetition (and not as the poorer version) of the past' (Lenclud 2003). From this perspective, fantasies of a lost Golden Age make no sense, while the view of everything as a circle engenders a greater sense of responsibility for our own actions – as Betty Laverdure of the Ojibwa said: 'It will come back.' Breaking the circle may impair linear history too. The Kyrgyz writer Chingiz Aitmatov (Aitmatow 1995) tells of a boy who had two stories – one, a gift from his grandfather, about the white hind from which the *ail* (local community) descended, the other, created by himself, about shape-shifting to meet his father who was returning on the white steamer. When a drunken uncle from outside the *ail* forces the grandfather to kill a white hind, thereby destroying the circle of clan history, the boy in his despair turns into a fish to swim towards Lake Issyk-Kul where he expects to meet the father on his steamer. Or so it seems – and the voice of the narrator ends the story in ambiguity. Continuity is, of course, not synonymous with perennial sameness, and change is a condition of continuity. Consider a tree, a river, a human being – they are never exactly the same and yet they remain recognisable over time. Change is primarily generated from within, and therefore strengthening their capacity to regenerate from within may help preserve their ability to change. That ability is endangered – perhaps terminated – by any or a combination of a number of factors interfering with continuity: sickness or violence, which may in turn be accidental or deliberate. In the world of discourse, violence may also be conceptual. Much of this has been inflicted on the folk over the years.

The folk and the populace

In 1937, Bertolt Brecht suggested that the term *Volk* (folk) be replaced by *Bevölkerung* (populace) on the grounds that 'those who make this change "already refuse to endorse many lies"' (Bausinger 1990: 1). Pointing to the historical connections with Herder and the Romantics, and to the significance of Riehl in particular for the development of the subject, Hermann Bausinger argued that ethnologists had good reasons to persist in using the term 'folk'. But since the 1960s the folk have been progressively ousted from academic discourse, despite some brave moves to the contrary, such as the Institute of Dialect and Folk Life Studies at Leeds (1960–84) or the attempt in the 1980s to talk into existence a discipline of 'folk geography' that claimed Sigurd Erixson and Estyn Evans as its ancestors and the Institute of Irish Studies at Queen's University, Belfast, as its stronghold (Lornell and Mealor 1983: 53). The popular revival of 'folk music' notwithstanding, in academic terms the folk as a contemporary category have been declared irrelevant, primarily on the grounds that, with industrialisation, modernisation, globalisation and all that, they have already as good as disappeared from our industrialised, modernised, globalised contemporary North Atlantic culture and society. They now only exist in countries that are far away and suitably foreign, such as Central Asia, or the past we call our heritage.

Every now and then, a voice in the wilderness claims that the folk have not gone away, but that we might have had the discursive wool pulled over our eyes – Bausinger said something to that effect in 1961, as did Hardt and Negri (2000), coming from a different angle, nearly 40 years later. Bausinger's point was that an overemphasis on a disappearing peasantry may well have blinded us for the persistence of *Folk Culture in a World of Technology* and he supported that argument with a range of empirical examples from urban contexts. Hardt and Negri, political philosophers with an alleged tendency towards activism, see in the anti-globalisation and other protest movements a resurgence of what they call the 'multitude', a concept of the folk as a force that will not be swept away by the nation, the state or big business – nor by any fashionable academic discourse, for that matter. They do not see the folk as unified and cohesive but rather as pluralistic, flexible and capable of creating structures spontaneously. This is different from the concept that has replaced 'the folk' in social science research, where statistical generalisation and a focus on mass cultural phenomena have turned the spotlight onto an anodyne 'populace' (*die Masse*) moved about by metaphysical forces of the secular age, such as the market, fashion or simply globalisation. By contrast, folk as 'multitude' (*die Menge*) emphasises individual agency in the creation of collective identities and the places they relate to.

Place-identity in postmodernity

Globalisation is supposed to have disposed of the folk once and for all. However, in his history of the Iona Community, Ron Ferguson argues that 'the practicalities of every-day living ... [can foster] ... a healthy scepticism about global ideology' (Ferguson 1988: 203). Another factor in the disappearance of the folk has been the much-trumpeted end of history. 'What is important to Native communities is how the unique event may affect *their own* lives, not the global understanding of history' (Kidwell and Velie 2005: 42; emphasis added). It is easy to lose sight of this existential groundedness. Ferguson warns of the temptation 'to throw over history, to declare it redundant, and to talk only in the present tense' (Ferguson 1988: 195). But that is precisely what has happened with the advent of postmodernity. 'The logics of universalism ... modernization and globalization have sought to represent localized identities as historical, regressive characteristics, and have worked to undermine the old allegiances of place and community' (Morley and Robins 1995: viii). The problems arising with older notions of place identity are often linked to what David Harvey (1989) called 'space-time compression' – the substitution of pace for place. 'Place-identity ... becomes an important issue, because ... how we individuate ourselves [vis-à-vis place] shapes identity. Furthermore, if no one "knows their place" in this shifting collage world, how can a secure social order be fashioned or sustained?' (Harvey 1989: 302). At the height of the postmodernist fad, Harvey's analysis cautioned against the fuzzy rootlessness that was being celebrated, and not just by academics. It also pointed back to the topological role of ethnology as indicated by Riehl and the tradition of 'general statistics'. The close ties of place identities with the political boundaries of nations postulated by the nineteenth-century nationalists have certainly loosened, if not dissolved, but 'they still resonate throughout the imaginations of communities' (Reisenleitner 2001: 9). At the end of the twentieth century there was much emphasis on spaces and flows but spaces provide no grounding for identities – places do. Spaces can be turned into places if they are endowed with meaning. Tradition constitutes a key element of such endowment. A defining characteristic of modernity has been the belief – if not the fear – that traditions everywhere are under threat of extinction. On the other hand, a paradox of our time is that local traditions are widely highlighted, not just for the purposes of heritage tourism (Tilley 2006: 12). The concept of 'tradition' has been under the discursive magnifying glass for some time, not just since Lord Giddens (1999) in his third Reith Lecture diagnosed a lack of engagement with the issue.

Invented traditions

Tradition is nowadays widely regarded as invented, and invented traditions are said to provide the foundations for many forms of hegemony and mechanisms of inclusion and exclusion (Reisenleitner 2001: 8). Introducing the book *The End of Tradition?*, the editor neatly sums up the common reading of Hobsbawm's and Ranger's influential collection, *The Invention of Tradition*, in terms of 'Hobsbawm's and Ranger's notion regarding all tradition as being invented' (AlSayyad 2004: 23). That claim is itself an invented tradition of academic discourse, which highlights the virtue of reading a text properly before quoting it. Apart from the fact that 'an invented tradition is still a tradition' (Gbadegesin 2007), Hobsbawm and Ranger actually said something different. In his introduction to the book, Hobsbawm writes that '"Traditions" which appear or claim to be old are often quite recent in origin and *sometimes* invented' (Hobsbawm 1992: 1; emphasis added). The problem with the debate about 'tradition' is that it uses the same term to refer to a state of affairs, a process and a product. This can only cause confusion. The Latin root of 'tradition', *tradere*, refers to the act of transmission rather than to the object, value or practice that is being transmitted. Hence we may rethink 'tradition' as the process of and necessary context for the transmission of knowledge and related practices across space and, in particular, through time. Tradition need not be primordial; it may be new, created even here and now, at any time. What characterises tradition is not its age but the fact that it is being passed on; this transmission is not dependent on kinship but may also happen, for example, through apprenticeship. Consequently, teaching and learning are processual traditions. What responsibilities arise if we understand them in this way? Is all teaching and learning necessarily a process of tradition? If not, which aspects are and how do we know the difference? A key aspect of tradition is that it has not simply been transmitted but that transmission happened in a particular manner (Lenclud 2003: 76). It is in this process that change occurs as knowledge and its expression in practices are handed on. While customs and tradition may be idealised as invariable, both live by subverting that ideal within certain limits. When they cease to be variable, they either disappear or become fossilised as 'heritage'.

If tradition is identified as invented, that is, made up, not real, not authentic, then that identification presupposes the existence of a not made up, real, authentic tradition to which it can be compared. Therefore, it may be argued from a folk point of view that the proponents of a wholesale 'invention of tradition' argument are actually the ones who are essentialising 'tradition', and that their argument depends on this essentialisation far more than the world view they set out to demolish. Bourgeois nostalgia is often cited as a main factor in the nineteenth-century invention of traditions (e.g. Sennett 1996), but the bourgeoisie had different reasons than the folk to look towards the past. For both, history provided legitimacy for the present but the folk valued tradition out of 'a pragmatic attitude toward sure and well-tried knowledge'. Peasants usually were interested in their ancestors 'for concrete economic reasons: traditions of farm ownership, land rights, and the like' (Löfgren 1987: 33). This pragmatism reflects 'the security based on *simplicity* in the face of a complicated reality' that, according to Bausinger, 'characterizes the folk culture as a whole' (Bausinger 1990: 115). This is a particular kind of simplicity and as ethnologists we need to be acutely aware of the danger arising for folk culture when we simplify its complexity 'in a short-circuiting, inadequate manner and by putting the stress in the wrong places'. Bausinger's warning had the past practices of ethnology in mind, but it was also more than a little prophetic. Much of the discourse over the folk and their place in recent decades, informed by visions of the uses and abuses of concepts of the folk in the past, has simplified, short-circuited and misplaced the emphasis in order to exorcise the folk from their place, thereby destroying the historical and ecological unity of places and the folk who make them.

The destruction of place

For Native Americans, the environment physically manifests their spirituality. Moreover, landscape carries moral implications. Events associated with certain places hold current meaning for contemporary Indians even though they may have happened a long time ago – places have lives of their own (Nabokov 2007). The philosopher Edward Casey notes that during the age of exploration 'the domination of native peoples was accomplished by their deplacialization: the systematic destruction of regional landscapes that served as the concrete settings for local cultures' (Casey 1998: 77). The destruction of places went hand in hand with the – quite literally: progressive – expulsion of the folk who were perceived as out of place in this modern world. In the 'new world' of discovery, those who survived ended up in reservations or shanty towns, while in the 'old world' they found themselves relegated to marginal and peripheral areas. Some provide a folksy dash of colour – in the picturesque *Gaeltacht*, Santa's Lapland or the Scottish Highlands – while others suffer in the sink estates of our metropolitan centres, or in folk singer George Papavgeris' 'Anytown' – reminders of Bausinger's (1990) insistence that folk places are by no means only rural.

The substitution of spaces for places culminated in the space-time compression diagnosed by Harvey (1989) and finds expression in academic books like *Ireland: Space, Text, Time* (Harte et al. 2005). Not enough that sites of memory – *lieux de mémoire* – have supplanted ecosystems of memory – *milieux de mémoire* – as Pierre Nora observed (cited in Morley and Robins 1995: 87): text has taken over from lived experience. Surfing the constructivist tsunami, the critics of tradition have cheerfully mingled words and concepts to a point where nothing made by anyone can be considered authentic because it is 'invented' and therefore everything (and everybody) becomes disposable. One may ask whose interests are best served by this ideology but that is a debate for another day. A constructivist approach seriously disregards people's ecological relationships. In a world seen purely as a social or cultural construct, the environment can be seen simply in terms of resources at hand, ready for exploitation (Frykman and Gilje 2003: 11). There are many today, not only in post-Communist Europe, who fear the market and its particular mechanisms of globalisa-tion, which seem to have the power to destroy what totalitarian systems were unable to extinguish (Venclova 2001: 77). In this regard, the poet Thomas Venclova speaks of two cultures – the old culture of memory and the new culture of forgetting. And despite appearances, the latter can be uncannily linked to the celebration of heritage. When corrupt and cruel utopias extinguish memory and denigrate people, the cult of remembrance becomes paralysed (Venclova 2001: 78) – and associated traditions become 'heritage' (Kockel 2007c). But that, too, is a topic for another day.

Putting the folk in their place

Venclova (2001: 76–7) wonders whether at least some parts of the culture of his own and numerous other regions will be reborn in the twenty-first century. This is by no means a romantic hankering for a golden age when all was bright and beautiful. Rather, he refers to respect for a collective sense of cultural values and appropriate actions engender-ing a sense of responsibility towards the 'community' as Native American scholars use the term (e.g. Fixico 2003). A key aspect of such community is that its members 'know their place' – not in a 'do-as-you-are-told' sense but as a matter of ecological responsibility that is a defining element of their membership. Of course, this does not imply an easy solution to contemporary environmental problems – just put the folk in their place and all shall be well! Anthropologists know that 'human beings have no "natural" propensity for living sustainably with their environment', and identifying cultures that relate sensitively to their ecological context is 'not as easy as pointing to non-industrial peoples' (Milton 1996: 222) – not even to the much-cited Native Americans. Kay Milton (1996: 223) argues that we may be able to create 'sustainable ways of living out of bits and pieces selected from diverse cultures, but it would be unwise to attempt this without first understanding them in their original contexts, and appreciating the consequences of taking them out of those contexts'.

A century ago, Patrick Geddes coined the triad of 'Folk – Work – Place', using the term 'folk' to situate the individual in culture and community. 'Place', in this framework, is not just a topographic location but 'a "Work-Place" of productive activity and a "Folk-Place" of residences' (Law 2005: 6–7). It is a specific locality 'identified by the community and the social relations that are played out in that place ... [and] ... imbued with meaning from history and tradition' (Moxnes 2003: 12). Instead of seeing a 'place' as a fixed, bound 'site of normative authenticity', we may see its identity as 'formed in interactions with the outside and with others' (Massey 1994: 168–9). Developing an integrated, ecological concept of place as a location of mutual householding – what the theologian Halvor Moxnes (2003: 157) calls 'generalized reciprocity' – means 'turning away from domination and control of place' that have characterised place politics, and towards 'appropriation and use of place'. With apologies to the theologian for adapting his Christological argument (cf. Moxnes 2003: 12) for my purposes, I propose that we do not try 'to find the right place for [the folk], as if there were only one fixed position and category to put [them] into. Rather, we will see [the folk] and [their] activity as engaged in a contest over places'. This is only the first step. If we want to be good ethnologists and discover the roots of the folk in place, we have to become radical (from the Latin: *radix*, the root). But, as Bausinger (1990: 115) asked many years ago with an eye on the history of this field of enquiry, 'are all such natural metaphors and the concomitant notions best renounced entirely?' A full answer will have to wait for another day, but let me suggest, following Ulf Hannerz (1996), that we do need to rethink some of the keywords of our trade, and in some instances dust them off and critically rehabilitate them.

The field where anthropologists do their fieldwork used to be 'a rather fixed entity ... some place you could get to know by covering it on foot and engaging with its people face to face. And it used to be self-evidently a matter of "being there" – away, rather than "here"' (Hannerz 2006: 23). While social anthropologists 'devised techniques for getting *into* a new culture, European ethnologists ... struggled with the problem of getting *out*, of distancing themselves from far-too-familiar surroundings' (Löfgren and Frykman 1987: 4; original emphasis). Both deal with situations where the creation of hierarchies, although (usually) contrary to the researcher's intention, is unavoidable. As even the earliest collectors of folk culture knew, by pointing 'to the existence of a category of folk, you distance yourself from the category' (Hylland 2001: 24). Moreover, what Ron Ferguson said about the Church is equally true for academia: many folk see it 'not as a fellowship of liberated and accepting people, but as a group of well-heeled, judgemental people who have been fortunate enough to have got their lives together' (Ferguson 1988: 175). The interests of academics are often perceived as removed from those of local folk. 'Local knowledge' means the collective knowledge of a particular community, including their oral traditions, shared memory and deep familiarity with their environment (Nabokov 2007). Academics, however, usually seek some kind of 'global knowledge' that is valid universally and at all times. It would help a lot if we could bring ourselves to acknowledge that identities are *lived* as well as *constructed*, and that the folk are 'active agents in their history, not simply passive victims or obstacles to someone else's progress' (Kidwell and Velie 2005: 42).

For all these reasons, anyone interested in and concerned for the folk must make a critical distinction: 'between wanting to extract by means of collection something from the folk and focusing upon giving something to the folk' (Hylland 2001: 20). To what extent, then, should we submit our research to the folk for scrutiny and authorisation? Should the folk determine our research agendas to meet the needs of their communities? This raises the tricky issue of 'who speaks for communities where factional disputes may pit members against each other' (Kidwell and Velie 2005: xii). We need to balance engagement and analysis. As participants in other people's lives, scholars 'should use imagination and intuition, allowing themselves to be inspired and implicated by the specific situation' (Frykman and Gilje 2003: 10). Ethnologists are not looking for generalisations, but try to 'describe the specific that is deeply experienced and is therefore universal ... and by the same token ... deeply communal in the singular. Naturally, such ideas point in the direction of an ecological awareness. ... [I]dentity is worked out in relation to an existing environment, to objects and to places' (Frykman and Gilje 2003: 10–11).

Ethnology is about specifics and specificity, about the folk in the places they inhabit. Universals may be derived from its material by careful comparative study but the main focus is always first on the specific and then on the general. That does not mean that ethnology is purely inductive – although at times in its past it was seen as such – but it sets ethnology somewhat apart from other, more mainstream social sciences and humanities disciplines that aim at universals. One might perhaps say that the key universal in ethnology is that there are few universals or, to put it differently, that between essentially fundamentalist claims to a 'universal truth' and the postmodernist free-for-all of an essentially non-identity politics, the ethnological challenge is to find the courage to postulate the legitimacy of actually existing differences, sensitive approaches to exploring their actual foundations and appropriate ways of dealing politically with their practical implications at a societal level.

Turning the world upside down?

With apologies to yet another theologian for adapting his argument, if we are genuinely concerned to give something back to the folk, then '[t]he task of ... equipping ... the [folk] ... must start where the [folk] are, and not where [we] would like them to be' (Ferguson 1988: 201). Speaking from experience as an activist, Alastair McIntosh (1998) suggests that 'work with communities involves a three-part process of re-membering what has been lost, re-visioning how we could live, and re-claiming what is needed to achieve that vision'. One danger for activists is that they are 'co-opted by a benign establishment'; on the other hand, 'radical and innovative ideas, discussed by enthusiasts in an intense atmosphere which is light years away from the concerns of ordinary people, can do strange things to otherwise nice people' (Ferguson 1988: 202).

Like Ron Ferguson's Christianity, ethnology 'is at its best when it is radical, freewheeling, iconoclastic, prophetic, refusing to bend the knee. When it degenerates into a prudential buttress for the powers-that-be, it sells [the folk] down the river'. But Ferguson cautions: 'It is ... insulting and dehumanizing to romanticize the poor and to present the oppressed as innocent refugees from Eden, as many guilt-ridden middle-class radicals do' (Ferguson 1988: 175–6). Whatever our roots, by becoming academics we have also, incurably, become middle class and joined the establishment. The folk know that, even if we don't. Being aware of this insurmountable distance is the first step towards turning the tensions arising from it into creative energy. While it will not propel us back to any Eden, that energy, used sensitively and responsibly, can help us jointly to re-member ourselves in community, re-vision the earth as a 'common treasury' and re-claim simple places – not just for some folk, but for all of us.

6
Fifth Journey – Towards Castalia: *To Re-Place Europe*

The scholarly province of Castalia is one of the great non-places in European literature. Yet the real utopia is the Castalia that will come after Castalia. In the final scene of Hermann Hesse's 1943 novel, *Das Glasperlenspiel*, the Magister Ludi, Joseph Knecht, having left the ivory tower to teach and learn in the everyday world of other provinces, drowns in a mountain lake where he is swimming with his student Tito.

According to Bill Readings, we are living in an age where universities are losing their status as institutions for the education of a society constituted as a nation state and are trying to behave more like transnational corporations. In the US at least, societal relevance of knowledge production and its contribution to the education of citizens are making way for a concern with performance indicators; the conditions of academic work are such that the pressure to produce undermines or overshadows the dialogue about societal relevance (Bendix 1999: 100). The Ivory Tower is being replaced with an Iron Cage, by the sound of it. Well, that may be in the US; it could surely never happen in Europe – or so we might have thought when Regina Bendix reviewed the American experience at the end of the twentieth century. A decade later, we are perhaps a little wiser, in any case a little less naïve. While few of us would quibble with the 'pursuit of excellence' that figures prominently in the new rhetoric, 'excellence' is a double-edged (s)word, as Irène Bellier pointed out at a 2008 symposium of Europeanist anthropologists in Madrid: across the disciplines, many researchers regarded as 'excellent' in the past have actually caused major harm to other human beings and indeed the planet.

In Chapter 5, I started exploring the specificity of European ethnology and the societal relevance of the knowledge it produces, at this juncture and perhaps beyond. Ethnology involves translations between different living conditions (Bendix 1999: 105). According to Claude Lévi-Strauss, being a nonconformist and rejecting one's own society are virtual preconditions for a career in ethnology. Not everybody would go that far. Justin Stagl, for example, sees Georg Simmel's concept of the restlessly wandering stranger or Robert Park's 'marginal man' as more accurate representations. The combination of distance and nearness, indifference and engagement makes the ethnologist flexible, and suitable as a mediator (Burckhardt-Seebass 1999: 121–2). There is hardly another subject where living and working are as closely connected as in European ethnology – hence the strong reaction against approaches that engage with representations, but not with primary research or personal experience, where everyday life is only recognisable in 'typographic traces' (Streng and Bakay 1999: 131). In many ways, the postmodern debate has been good for us; it challenged our styles of writing and self-reflexivity and demolished some cherished concepts. At the same time, the discussion of postmodern identities has placed too much emphasis on 'post' processes like the disintegration of identities, or the fragmentation of just about everything, from the family to the nation state. The diagnosis depends, of course, not only on where you are looking but also where you are looking from (Löfgren 2001: 153) and how that looking is done.

The subjectivist turn in European ethnology during the 1970s may have been merely a poorly disguised attempt to give in to curiosity and, for example by using survey techniques, to obtain 'results' quickly and cheaply – 'Fast Food' research, as Martin Scharfe (2001: 68) has called this approach recently. Instant gratification is the order of the day. There is nothing new in this. Hermann Bausinger, in his 1961 book, *Volkskultur in der technischen Welt*, described instant availability as a cultural principle of everyday life nearly half a century ago, in a work that proved seminal for European ethnology. Quick-fix surveys of attitudes and opinions have their uses, but they are no substitute for direct engagement with lived experience, for intensive and extensive fieldwork as the foundation of ethnographic and ethnological research.

Quantum physicists and European ethnologists know that the researcher's presence in the field influences the research process and the findings. Unlike the physicist, however, the ethnologist engages in the cultural representation of other social worlds, and this creates not just abstract epistemological problems but concrete political ones too. How we tell the stories that we actually can tell is more than a playful question; it is a matter of political responsibility (Niedermüller 1999: 64).

Löfgren (2001: 153) identifies what he describes as the five ethnological virtues – 'what we are good at':

- a historical perspective and a comparative approach
- an interest in everyday life and its materiality
- the ethnographic approach and its moving searchlight
- the focus on culture in context (the importance of contextualising, concretising)
- the role of the *bricoleur* in search of theory and methods

It is one of the strengths of European ethnology that, due to its diverse interdisciplinary connections, it is particularly open for multidisciplinary collaboration. Stressing the field's characteristic perspective on experience, practice and everyday culture within the broad framework of historically oriented cultural research serves to maintain the interdisciplinary location and distinctive contribution of European ethnology (Hengartner 2001: 46).

At this point it is worth noting that the terms 'multidisciplinary' and 'interdisciplinary' are frequently used as synonyms, and this muddle has seriously clouded the vision of university course directors and research assessors alike. Most faculties are multidisciplinary environments where practitioners of different disciplines may engage in mutually enriching collaborative research projects or teach joint modules, thus broadening their students' horizon at least as an aspiration. By contrast, genuinely interdisciplinary work is located *between* the disciplines – not as a postmodern 'pick-and-mix' ragbag or a purely pragmatic combination of useful elements from different disciplines but with a philosophically grounded epistemology and methodology reflecting its research foci. On the research side, this means that genuinely interdisciplinary work is difficult to place in mainstream disciplinary journals because it does not play to the canon of any one discipline. In teaching, it similarly challenges student expectations of a straight and narrow path to a degree, demanding instead engagement with a broader range of theories and methods than traditional or even multidisciplinary programmes. It also requires something that has increasingly fallen out of fashion as academic teaching, and research has become more and more instrumentalised – a deep engagement with first principles of philosophy. Only a solid grounding in logic, argument and evidence can provide foundations for sound interdisciplinary epistemologies. In the language of contemporary spin, these concerns may even be marketable as 'transferable skills', even if that term nowadays more commonly refers to the pressing of buttons and the flicking of switches.

Liberating the ethnological imagination[1]

The sub-theme for the 2008 congress of the *Société Internationale d'Ethnologie et de Folklore* (SIEF) was 'Liberating the Ethnological Imagination'. The implications of this are at least threefold:

1. that there is an ethnological imagination, and therefore ethnology is creative, not simply an unimaginative gathering of 'facts';
2. that this imagination is currently in a state of captivity (as Rousseau might have said: 'born free, but everywhere in chains'), preventing it from unfolding its creative potential and
3. that there are ways and means of breaking out of this captivity.

One might add a fourth implication, namely that such a jail breaking would be a good thing to achieve. This is by no means as self-evident as it might seem to some – there are prisoners who prefer the shelter of guarded routines to the vagaries of the world 'outside'.

Towards a *Daseins*-ethnology

Over the years, European ethnology has become highly adept at reinventing itself. This is not the place and occasion to revisit the various incarnations, some of which are discussed in a recent book (Nic Craith, Kockel and Johler 2008). Instead, I want to consider the challenge posed in the title of the 2008 SIEF conference. What does that actually mean: 'liberating the ethnological imagination'? What are the sources this liberation may feed on? What could it look like in practice? What (and who) makes imaginative ethnology and who benefits from it? Is ethnology worth the trouble, or should we just resign ourselves to being appendages of larger units for teaching and research purposes?

One question not asked here before now might be regarded as rather crucial: What is ethnology? In the 1980s, geography underwent an identity crisis during which many prominent practitioners claimed that 'geography is what geographers do'. As a doctoral student I smiled at this and thought it a smart cop-out. Then I witnessed anthropology going down the same route. And, of course, European ethnology has been there at least since the Falkenstein symposium (Kockel 1999a). In one sense the statement is true: ethnology is what ethnologists do. But there are many other senses. European ethnologists do history, sociology, geography, political economy, literature, art, architecture and so on. It might therefore be more accurate to say that 'European ethnology is how European ethnologists do things.' The problem with this is that you will have 'real' historians, sociologists and so on who claim from their disciplinary high horses that European ethnologists lack the 'proper' disciplinary rigour – which does have a grain of truth in it: European ethnologists can indeed be undisciplined academics. And Foucault tells us what happens to undisciplined members of any society: incarceration of one sort or another. There are subtle ways of incarceration – the creation of an audit culture, which inevitably stifles smaller subjects more than larger ones that have a bigger staff to whom tasks may be delegated, is only one aspect. Then there is – still, after all these years – Max Weber's *stahlhartes Gehäuse* (famously rendered by Talcott Parsons as the 'iron cage'), an encasement as hard as steel into which a rampant capitalism inescapably straps its subjects. Third, there are snares set by some past preoccupations of European ethnology, both methodically and in terms of subject matter, which may still be vigorously defended as cherished 'traditions' when in fact they have long become fossilised and devoid of the dynamic characterising genuine traditions. With the cat now firmly among the pigeons, let me return to the questions raised earlier. What I am offering here are by no means answers in the sense of any philosophical truths – empirical, analytical or otherwise – but tentative interpretations from a personal perspective: a vision that may be one among many. At this stage in the debate to do otherwise would only mean incarcerating the imagination once more.

Thomas Højrup (2003: 2) identifies a 'cultural-relational dialectic' that conditions ethnology: 'our concepts and values are a product of cultural life-modes' while they also 'determine the kinds of life-modes we can conceive'. This leads to an important insight: 'Ethnocentrism and *the continuing effort to transcend ethnocentrism* are therefore fundamental features of ethnology' (original emphasis) that help us understand different life-modes and the relations between them. Ethnology, in its continuous effort to transcend ethnocentrism, needs to study the foundations of ethnocentrism rather than merely dismiss it as an uncomfortable heritage. This includes the courage to difference evoked many years ago by Werner Schiffauer (1996), who argued that anthropology ought to overcome its 'fear of difference'. Since the proclaimed advent of postmodernity, many disciplines have indeed become afraid of postulating cultural difference. European ethnology should stand up and speak out against this dangerous orientation. The aim is not an assertion of difference as superiority but reclamation of a spirit of appreciation of difference and diversity, regarding these not just as elements of ad-lib performances (as postmodernists do) but as characterising the everyday life of groups and individuals, thus allowing people to be different and enjoy this diversity without having to pretend it is merely some kind of mock difference put on for the sake of carnival or other purposes of entertainment.

The critique has been joined by some sociologists coming from and working within intercultural contexts. In his 'cross-cultural critique of modernity', Fuyuki Kurasawa reads classic authorities of his trade as representing an ideological counter-current contesting the social order of Western modernity. Rousseau, Marx, Durkheim, Lévi-Strauss, Foucault and even Max Weber are called upon as witnesses to the existence of what Kurasawa terms 'the ethnological imagination'.[2] 'Ethnological', for Kurasawa (2004: 12), designates 'in the broad and etymologically literal sense ... the comparative study of societies aiming to produce critical interpretations of the modern West'. He quotes (loc. cit.) Merleau-Ponty (1960: 150) who sees ethnology not as

> a speciality defined by a particular object ... [but as] ... a way of thinking, one which imposes itself when the object is 'other,' and demands that we transform ourselves. Thus we become the ethnologists of our own society if we distance ourselves from it.

With the term 'imagination', Kurasawa (2004: 12) seeks to highlight 'the mythical character of constructs of otherness found in cross-cultural reflection'. These constructs are myths in that they represent 'related sets of beliefs and values created to rhetorically explain what Euro-American societies have become in relation to their pasts and their futures' (op. cit.: 13). The 'ethnological imagination' produces 'a critical examination of this sociohistorical formation from a distance and through a comparative perspective acquired by way of encounters with widely differing ways of being in the world' (loc. cit.). Kurasawa therefore challenges both the fashionable dismissal of social theory as imperialist and ethnocentric and the common denial of the intercultural basis for much of the disciplinary canon. Thus attacking the twin giants of universalism and particularism, he suggests that, by cultivating the ethnological imagination in an increasingly multicultural world, we can enable social theory to respond better to issues of identity and boundaries, not just at the level of empirical detail but also analytically, with regard to 'the West' and 'modernity'. One might say that this is all very well for Kurasawa's discipline of sociology, which would benefit from some ethnological imagination. However, I think there is food for thought here beyond that, not least in his use of Merleau-Ponty's definition of ethnology – which, coming from the French, embraces social and cultural anthropology along with European ethnology in most of its various guises – and the hint of a Heideggerian framework, which is also expressed in the reference to *Dasein* [being there] woven into the title of a collection of essays on phenomenological approaches to the analysis of culture by European ethnologists and anthropologists (Frykman and Gilje 2003).

By virtue of its name, European ethnology is perhaps more liable than most other fields of research to be charged with the sin of Eurocentrism. Before we (over)react to this charge, we ought to remind ourselves and our critics that Eurocentrism is just one form of ethnocentrism and that, as such, it constitutes a legitimate and, indeed, necessary subject for examination, as Thomas Højrup suggests. The postmodernist response to the problem has long been to declare Europe a delusion, thus making Europeans non-existent by definition. The 'folk' – with which earlier incarnations of European ethnology have been so eagerly concerned – have been ousted, replaced by an anodyne populace. The latter implies sameness flavoured with some identity-warehouse colouring. 'Identities' projected in this way are fleeting, forever changing and unstructured. The celebration of these effectively 'indifferent' identities plays into the hands of a closet form of fascism where being different invariably means deviant, and therefore a legitimate target for ostracising. But what should we do about that? As European ethnologists, given the past of our field, how are we going to celebrate difference without once again playing into the hands of the perpetrators of 'blood and soil'? How do we generate new terms that allow us to revisit old concerns free from historical baggage?

In Chapter 5, I have suggested that we might go one step further and grab the European (ethnological) bull by the horns, wrestling with a new critical understanding of indigeneity in the European context. Could we take a cue (or at least a clue) from Native American studies? This would not be a matter of reading their culture through our categories – such as property rights – or vice versa; nor would it be about learning through communication between different cultures. Instead, like the autobiography of Black Hawk (Pratt 2001: 109), it would be about 'ways of seeing and understanding the place that sustained the life' of the people of Europe. Unfortunately, the discourse of 'nativeness' has been usurped by the political Right for xenophobic ends, and Europe has lost its indigeneity. An element of indigeneity may be visible in the

Central European *tuteishyi*, 'those who are simply "from here," even if that "here" changes in relation to the "theres" which have shaped and defined the territory' (Ivakhiv 2006: 38–9):

> The *tuteishyi* represents ... a person ... who is uncertain as to whether s/he is a nationality, ethnicity, or part of some other substance (religious denomination, et al.), but who is defined by the place in which s/he remains (and moves) while empires, armies, time-zones, and global economic forces move in and out of range ... rooted enough in his or her own space (Tarasiewicz's forests, Maszlanko's fields), mobile in the tracks and paths carved out through earthy meanderings in the interstices of nations and empires.

Could supranational bodies like the EU help empower these indigenous Europeans? And what could the role of European ethnology be in the process? After many years of soul-searching and reconstruction, European ethnology's focus on certain keywords – such as culture, everyday, historicity, identity (Bausinger et al. 1993) – remains and, combined with its methodological pluralism, uniquely equips its practitioners to address problems associated with recovering indigeneity, in Europe and elsewhere.

However, this cannot, must not, be 'salvage ethnology' in the service of a colonial project – internal or overseas – as we saw it in the nineteenth and twentieth centuries, but instead contribute to the subversive emancipation of the folk as postulated, for example, by the 'progressive patriot' singer and songwriter Billy Bragg (2007: 13) who rejects the 'rituals of pomp and circumstance ... designed to detract attention from the iniquities of the present by constant reference to a more glorious past'.

As a niche subject coming from the sidelines, European ethnology will hardly be able to conquer academia, by storm or otherwise. The scattering of graduates has ensured that there are European ethnologists working in many more universities and other research institutions than just those that offer a department or institute for this kind of work, under whatever title may be fashionable or locally acceptable. There will be obvious pressures to assimilate, to blend into whatever disciplinary teaching of undergraduate students in particular butters our bread. Where we have an institutional base, the prospect of a merger and takeover is always on the horizon. This could make anyone despondent. But it should not. And it need not, if we can liberate our own ethnological imagination a bit.

Insistence on the purification and maintenance of 'our own' disciplinary canon will seal the fate of European ethnology and consign it as an artefact to the Museum of Ideas That Have Had Their Day. Like those of the *tuteishyi*, the roots of European ethnology may well be strong but they are certainly not pure. That makes our field particularly suitable for interdisciplinary work. I would even claim that its concerns and methodology put it at the leading edge of interdisciplinarity (Kockel 2009). This is our strength, and we should play to it. As a small field, we pose no threat to other disciplines and research fields, but we have much to offer them. Mutual enrichment can flow from greater engagement with some fields in particular: the creative and performing arts, including fine art and digital media; creative writing, especially poetry; and human ecology. This is not an exhaustive list, nor should it be taken as exclusive of areas not mentioned – far from it. I must also confess to a certain bias arising from the fact that these are the areas my colleagues and I at the University of Ulster are most actively involved with. But liberating the ethnological imagination will take a bit of time and effort, and you have to start somewhere.

Towards an ethno-anthropology of Europe

When the first issue of the *Anthropological Journal on European Cultures* (AJEC) was published in 1990 by the European Centre for Traditional and Regional Cultures (ECTARC), Europe was a different place. As the director of ECTARC at the time, Franz-Josef Stummann (1990: 7), explained in his introduction to that issue, the 'magical date of 1992', heralding the Single European Market as a significant step towards European integration, had 'a substantial bearing' on the foundation of the journal. Moreover, the Berlin Wall, symbol of the political divide that cut right through Cold War Europe, had crumbled the previous year. German unification was imminent but little else seemed predictable. By the time the journal was relaunched in 2008, 18 years and two Gulf Wars later, not only had the EU acquired 15 new member states, ten of them former Communist countries, but we also had been told to perceive a new divide – between a 'new' Europe and an 'old' one.

Simplistically coined in the context of the second Gulf War, this distinction is also reflected in the discourses that have emerged since the late 1980s, pitching a commodity view of the world – as in the concept of 'cultural industries' or the almost universal marriage of cultural studies and business training under the label 'European Studies' – against a critical, more differentiated view that, for much of the past two decades, seems to have been on the retreat. In these circumstances, Stummann's

(1990: 9) remark that concepts of 'culture and cultural dynamics need the interdisciplinary orientated anthropologist who maintains the ability to distinguish and to differentiate' is as valid today as it was then.

In the first research article in AJEC, founding editor Ina-Maria Greverus developed a critical perspective on what she described as 'a growing postmodern indifference to the utopian Not-Yet' (Greverus 1990: 14). Discussing an interdisciplinary conference organised by the *Westdeutsche Rektorenkonferenz* (West German consortium of presidents of universities and colleges) in 1988, she noted a rather 'schizophrenic break' in 'the conviction manifest at the conference that the apparent ease with which we "already" move across boundaries within and between university disciplines offered a substitute for that "Not-Yet" by which one might begin the serious work of overcoming boundaries in the reality beyond the ghetto of those disciplines', and diagnosed a 'lack of concern for real conditions and practices as they exist in ... societies and their subordinate institutions such as universities' (op. cit.: 15). Looking around the disciplinary ghettoes that continue to frame the context for much of university research in the early twenty-first century, one can still observe 'the almost manic way in which theoretical pluralism and a variety of methodologies [are] conjured up, supposedly in order to overcome the boundaries between academic disciplines' (loc. cit.) while universities seem to be run, then as now, 'by bureaucrats in alliance with conformists and ... "shallow practicists"' (op. cit.: 16). Against this spectre, Greverus issued a rallying call for 'the rare few who are actively seeking out the possibilities of a humanities of the Not-Yet ... to take a critical stance ... not so much by means of cultural pessimism but by expanding our horizons of knowledge' (loc. cit.). A key element of this process is the development of a 'reciprocal understanding of Self and other' that cannot be achieved by thinking alone but requires 'the practice of empirical research as the perceptual experience of otherness'. This 'manifesto for empirical ethno-anthropology' (op. cit.: 17) remains a cornerstone of the journal.

A second cornerstone is the commitment to the interdisciplinary expanding of our horizons of knowledge. Greverus (1990: 19) argued that anthropology as 'an empirical science ... has to be understood not as a discipline within the current organization of scholarship, but as an interest in knowledge [*Erkenntnisinteresse*] which extends beyond all particular disciplines'. Many who enjoy the comfort and conceptual safety of canonical disciplinary boxes would suspect such bold claims to be driven by the imperialistic desire of a self-styled 'umbrella discipline'. After playing the fashionable game of multidisciplinary collaboration, they prefer to return to their silos with the respective canon intact. But their fears are born out of a misreading of the anthropological project, which is not about establishing a meta-discipline to dominate the humanities and beyond but rather about the legitimacy of diverse modes of inquiry. It is not in any hierarchical 'above', but in that epistemological 'beyond all particular disciplines' that the 'Not-Yet' may be found. Greverus outlined the horizons of such knowledge in some detail. Two of these in particular I want to recall here.

The praxis-oriented horizon 'implies encouraging and assisting ... practical action toward the goal of transforming societal structures' (Greverus 1990: 25). This challenge of an applied anthropology goes beyond and sometimes against approaches to intercultural communication in the service of business or military interests, and while the status of 'applied anthropology' remains contested, these issues are today debated widely. Closely connected with the praxis-oriented horizon is the holistic-ecological one. Critical of an 'anthropocentric ecology ... dominated by economic rationality and ... quantitative, measurable differences', Greverus (1990: 26) postulated a human ecology that approaches 'ecological praxis not only via the horizon of material action but also via the horizon of understanding "intended meanings"'.

Along with empirical grounding and praxis-oriented, holistic-ecological interdisciplinarity, a third cornerstone of the journal is experimental writing, 'a renewed appreciation for the literary aspects of ethnographic textualization, for rhetoric, fiction, and subjectivity' based on 'an awareness of the historical contingency of different modes of writing' (Greverus 1990: 28). This appreciation arises not least from the 'Writing Culture' debate and the development of critical forms of 'native' anthropology that transcend dangerous Euro-(or any other)centrisms; both are crucial aspects of the attempts at 'liberating the ethnological imagination', expressed in the theme of the 2008 SIEF congress referred to earlier. Any journal established as 'a platform for anthropological research on and interpretations of both the potential of living European cultures and the restrictions to their actual and potential life' (Greverus 1990: 29) has to confront contemporary ideologies that cast any reference to 'Europe', 'European culture', and certainly to any plurality of 'cultures', as 'essentialising' and therefore a bad thing. The deconstruction of 'essentialisms' has been vitally important for our coming to terms with the unsavoury pasts of our nations and disciplines and there is no scope for complacency here. However, much of the contemporary debate over 'essentialism' amounts to a silencing of the European voices, regardless of whether these are 'of' Europe, 'in' Europe or coming from some other corner of the globe. This matter cannot be resolved here but it needs to be engaged with continuously, questioning the *Erkenntnisinteresse* that informs such debates, if we want to develop anthropological perspectives on European cultures that go beyond particular disciplines, experiment with different ethnographic genres or otherwise expand the horizons of anthropological knowledge towards a practically and ecologically inspired 'humanities of the Not-Yet'.

Searching for Europe in debatable lands

There is a common perception that the place we come from is what we call 'home' – or, as I prefer to say: *Heimat*. Yet this *Heimat* may also be located in a utopian future – a 'Not Yet' kind of place, to be brought into existence through the creative acts of a liberated humanity (Bloch 1978). This is not the *going* home of the postmodern individual who, like Anglo-European settlers in America, such as the Ulster-Scots (Chapter 2), is always on the move to somewhere else, 'seeking out the next horizon, finding Eden in some other locale and ultimately in glory above' (Deffenbaugh 2006: 5). It is more the *coming* home as understood by Native Americans who stress the need 'to achieve a very clear intuition of what it means to live with integrity right where they are'. This 'coming home' may be referred to as the transformation of the world into *Heimat* – investing a particular world version with patterns of meaning generating authentic belonging and, perhaps, even a sense of community that at once grounds and transcends any individual identity. In terms of the analytical framework outlined earlier, *Heimat* becomes a place that we are (habitually) used to, replete with markers of our habits. One could say that we can wrap ourselves in a place as if in a cloak (or habit) for protection against the elements. Let us look at some places that people have become wrapped up in, to see whether, and what kind of, 'Europe' may be found there.

The journeys of this book started in Ulster, one of the 'debatable lands' of historical as well as contemporary Europe. Originally, the term 'debatable lands' designated areas of disputed sovereignty along the Anglo-Scottish border. Used in the plural, 'debatable lands' refers to any part of the border held to be doubtful; in the singular it usually refers to the area in the west, between the rivers Esk and Sark, where the border agreed in 1552 is marked on modern maps as the 'Scots Dyke'. The term 'Scots Dyke' has been used in recent years in various geographical and metaphorical senses by writers and singers. Historically and culturally, this area has been every bit as much a heartland of

Celticity as the highlands and islands with their Norse Viking cultural overlay. Strathclyde and Cumbria may have been Brythonic rather than Goidelic, and may have become 'hybridised' sooner and/or differently than other parts of the islands. But that does not take from their significance for the image of a Celtic world that stretched across the Western seaboard of Europe. King Arthur may have been a Romanised Cumbrian P-Celt leading a band of Sarmatian horsemen against the Saxons, as a recent movie suggests, or he may have been a post-Roman mythic incarnation of the Q-Celt Fionn MacCumhail. Either way, we should not lose sight of the fact that *Keltoi* was a term liberally applied by the ancient Greeks to just about anyone living north of their own realm and who was not in the habit of speaking proper Greek. It seems more appropriate to think of 'non-Classical' Europe when we are looking for inspiration in 'the presuppositions of an old Europe' (Biggs n.d.: 71). This 'non-Classical' old Europe stretches eastward from the Celtic Fringe as far as Europe goes (at least) and includes those parts where, according to some accounts, King Arthur's horsemen hailed from.

Sarmatia

Sarmatia is one of the lost provinces of Europe; at one time it was its centre. Located between Lithuania, Belarus, Ukraine and Poland, Sarmatia saw Joseph Roth reach for his pen, Czesław Miłosz stroll around its markets and fairs, and the Singers and Brodskys pack their cases – if they had time to do so. It was the dreamland of Johannes Bobrowski, the wild realm where all peoples and religions could find their place, had not History ploughed under everything time and time again. Martin Pollack invited twenty-five writers to speak about Sarmatia, to remember it and to explore the fault lines of its landscape. The result is a compendium of a lost land, the rediscovery of which could give Europe a different, more open name.[3]

Inseparably connected, as the names cited in the blurb of a collection of essays (Pollack 2005) suggest, with the *shtetl* culture of eastern central Europe, Sarmatia was and is a wider region and a debatable land. Large parts of it are, in the vernacular language, simply called 'in the frontier' (*ukraina*). In the sixteenth century, 'the spatial visibility of Sarmatia became eclipsed by the cartographical discovery of Europe' (Briedis 2008: 53). On maps from that period, Sarmatia moves north, from the area of modern-day Romania in 1556 (cf. Briedis 2008: 18) to modern-day Lithuania in 1572 (cf. op. cit.: 40); a 1593 map depicts it as Prussia and the adjacent forest wildernesses to the east and south. Generally speaking,

Sarmatia coincides, more or less, with the area of the Polish-Lithuanian Commonwealth at its territorial peak, divided by Muscovy into a south-eastern or 'Asian' and a north-western or 'European' part. Speculations about Arthurian horsemen and the potential origin of the possibly Pictish queen Guinevere among Finno-Ugric tribes of Sarmatia – or indeed the steppes beyond – point to connections across old Europe which are as intriguing as their twentieth-century equivalents, from the German Youth Movement to the art of Joseph Beuys (Kockel 1995). However, by the seventeenth century,

> the mythological name of Europe had become inseparable from its geographical body: surrounded and safeguarded by water on three sides, Europe gradually dissolved into the vast landmass of Asia. ... At the beginning of the Enlightenment, Europe's continental distinction was firmly implanted in the minds of the educated elite of the western world. Sarmatia was the opposite: born in the minds of ancient scholars, it faded away into the realm of phantasmagoria with the passing of time. By the middle of the eighteenth century, Sarmatia retreated back to its mythological origins and left the map of Europe.
>
> (Briedis 2008: 53–4)

In the north of Sarmatia, contemporary Lithuania likes to emphasise the fact that it was the last country in Europe to be Christianised. The dates vary, but tend to fall about a 1000 years after the date St Patrick set foot in Ireland. When the Reformation arrived shortly afterwards through traders and colonial powers, it came to a people who had barely begun to adapt the new religion to their old practices. While Lithuania is a predominantly Catholic country – its Jewish population largely extinguished by the Holocaust and the Protestant colonisers departed after 1945 – there coexists under a thin veneer of modernity a considerably older Europe.

In 2004, the Franco-German TV company Arte commissioned film-makers from every country of the expanded EU to produce short films presenting their country's vision of Europe. Surprisingly many of the entries chose religious, even mythical themes and virtually all entries played with symbols. Most of them did so from a contemporary, post-modern perspective using conventional forms of exposition, which makes the films accessible for students who are trained in that mould. However, several cohorts of my students have been utterly confounded by the Lithuanian entry (Bartas 2004). The almost monochrome film shows four children interacting and dreaming with nature, history and place. A toad and a paper boat, both resiliently indestructible, also play a major part. Although I can understand the film intuitively, I find it almost impossible to explain to the students what it says about Europe, except that it represents an altogether different way of looking at the world.

There is another aspect to Sarmatia that I should mention before moving on. Some years ago, I brought back a CD from my first visit to Lithuania. It was a recording produced by the folk choir *Vorusnėlė* from near Klaipėda, the formerly German city of Memel.[4] Perhaps I should not have been surprised when I heard them play a tune from my childhood – but I was. Many years ago I had learnt it, and loved it, as a song about nature and young love: *Zogen einst fünf wilde Schwäne* (Five wild swans once flew). The image of the five birch trees standing on the banks of the river Memel (or Nemunas) is forever engraved in my memory. Later I had found out that this was an anti-war song, dating back at least to the Thirty Years' War in the seventeenth century, and probably beyond. The song with its strange harmonies had been popular with the rambling groups of the Youth Movement, both between the two world wars and after 1945. What the unexpected encounter with the tune on my way to work invoked in me was not so much the fond memory of a childhood favourite but the realisation that I had a bond with Sarmatia that reached deeper into cultural history than my personal life horizon.

After some reflexive searching, I now understand how this bond came about, but the puzzle remains. As far as I know, neither side of my family has any connection with Sarmatia; indeed, neither parent ever so much as visited the region. And yet both parents were deeply enchanted by the *Land der dunklen Wälder und kristallnen Seen* (land of the dark forests and crystal-clear lakes), as the opening line of the *Ostpreußen-Lied*, the regional anthem, describes the region. This much I remember, and while I did not recall any of the stories and songs that must have accompanied this experience, had seen no pictures of it and never had any conscious longing to go there – other than the curiosity of the incurable traveller – I would instantly recognise the landscape of north-western Sarmatia as familiar when I visited there for the first time. It was like coming home. Indeed, it felt much more like coming home than returning to any place that I could technically call home ever has done. While I had developed a sense of at-homeness away from my birth region before, not least on the Celtic Fringe in both Ireland and Scotland, in those cases I had arrived with a considerable store of advance knowledge, my imagination fuelled by pictures, songs and stories, and so ready to engage with the place consciously. With

Sarmatia, the sequence was uncannily reversed – as, following my first visit there, I started reading texts about and from the region, I recognised images and passages that recollect memories of *Dämmerstunde* (twilight hour), when around sunset from late autumn until Lent my mother would read stories or recite poems by candle light, ballads like *Die Frauen von Nidden* by Agnes Miegel (A. Schmidt 1994: 57–8). The second half of this ballad is a dialogue between seven women, survivors of the Black Death, and the Great Dune that rises between the village of Nidden/Nida and the Baltic Sea. Partly defiant, partly reconciled to the inevitable cycle of life in this place, the heart of the Curonian Spit, the women invite 'little Mother', as they call the Great Dune, to bury them – and the dune came and draped herself over them (*die Düne kam und deckte sie zu*).[5] I had completely forgotten the rest of the poem and its context for some 40 years or more, except for that last line. What does this tell us about oral tradition, especially where the process of tradition may happen away from the ecological context to which it refers? What does it tell us about old Europe? What does it tell us about ways in which we inhabit – we dwell in and on – our world? Before pursuing these issues any further, I want to turn to a European frontier that is not on the outer edge but right in the middle of this subcontinent – the debatable lands of the former Iron Curtain.

Zonenrand

In the very north of Bavaria lies the district of Coburg, joined to its mighty southern neighbour by one of the many plebiscites carried out across Europe after the First World War. Had it not been for that plebiscite, I more than likely would not be here: my father, a Social Democrat from near Dresden, ended up in American-occupied Bavaria at the end of the Second World War and never crossed the Iron Curtain after that. Much of my early life took place in or near what then used to be called the *Zonenrandgebiet*, that is, the area where the three joint zones of Allied occupation that made up West Germany bordered on the fourth zone, known as East Germany. Until 1973, the latter was officially referred to as 'Middle Germany', and our school books represented 'Germany in the borders of 1937' – that is, before the Nazi land grab started – and the lost territories in the east as 'currently under Polish/Soviet administration'. As I have only recently begun to think about this, I will not delve too deeply into the matter but I would like to highlight briefly some aspects that may be of particular interest here.

Despite the later image of steel fences, land mines, tank barriers and, of course, the Berlin Wall, the Iron Curtain was for the first third of its existence far less brutal in outward appearance. One abiding memory from early childhood – free of any wider context, as these memories tend to be – is of a visit to a café owned by a school friend of my mother's, located right on the border and aptly named *Grenzlandcafé*. As we were sitting on the terrace, my father decided to walk over to the barbed wire fence to chat with the farmer working on the other side, and mother called after him: 'You'll get yourself shot!' He didn't. But my curiosity was awakened: Why should anyone get themselves shot for chatting with someone on the other side of a fence?

Images of Sunday walkers gathering on hill tops with binoculars to gaze into the land no longer accessible to them sit alongside images of parcels passed across the fence, always in the same direction: east. And then the wild boar – pictured in the *Coburger Tageblatt* newspaper – that had one of his legs ripped off by a newly laid landmine. No more chats and no more parcels – only the lonesome groups with their binoculars on the hilltops. As the border was fortified, viewing towers began to spring up on the western side, civilised equivalents of the grim army watchtowers on the eastern side: the border became a tourist attraction, almost an economic asset. The *Zonenrandgebiet*, especially after 1961, was an area earmarked for special development funding. Towns cut off from their traditional hinterland struggled to keep economically afloat while the rest of West Germany was enjoying the fruits of the post-war 'economic miracle'. On the other side of the 'inner-German' border, there was a five-kilometre wide stretch of land called the *Sperrzone* (off-limits area); anyone living or working there needed a special permit to do so. In 1952 and 1962, the East German government undertook two operations, code-named, respectively, *Ungeziefer* (vermin) and *Kornblume* (corn flower), during which more than 10,000 inhabitants of the *Sperrzone* were expelled – in official parlance: 'resettled' – from their villages. Well into the 1970s many more of these villages, which were perceived as located too close to the border, were bulldozed.

The border, known as *Zonengrenze*, was the first element in a multilayered structure of debatable lands, encompassing the lost territories and 'language islands' in the east, which provided the reference framework for German identity post-1945 on either side of the Iron Curtain. The *Zonenrandgebiet* was not really part of the everyday post-war experience of the majority of Germans in the West, just like the *Sperrzone* on the other side was not really part of the everyday post-war East. They were, in an uncanny sort of way, negative exemplars of what Hermann Bausinger has called *Binnenexotik* – the exotic within. At the same time, these two areas were quintessential for the self-definition of the respective political system and, at a different level, both played a key role in the popular sense of belonging. Thus the *Zonengrenze* was a peculiar expression of that 'particular paradox of a liminality that *both* joins *and* divides' (Biggs n.d.: 18). Not surprisingly, when the border was finally removed, a number of border museums were created, some of them in situ, others by gathering buildings and other structures in a kind of theme park. The border itself, the landmines now cleared, is being turned into a nature reserve and ramblers' paradise – a Scots Dyke for at-home-less Germans.

In the late 1970s, a pocket of West Germany protruding into East Germany along the river Elbe became the site of the only attempt (so far) at secession from the Federal Republic of Germany. Surrounded on three sides by the Iron Curtain, the Wendland seemed an obvious place for the disposal of nuclear waste. The local population saw the matter differently and their ongoing struggle is well documented. On 3 May 1980, the *Freie Republik Wendland* was proclaimed by some 5000 local inhabitants supported by a 'rainbow' coalition of non-locals from all walks of life and all regions of West Germany. Although bulldozed by the authorities some weeks later, the Republic continues to exist as broadcasts, publications and other grass-roots political actions expressing opposition to the nuclear dump testify. Whether this was a genuine attempt at secession or merely a sophisticated political joke is a moot point. The Republic was a bold statement of local resistance to remote political control. For those not originally from the Wendland, it offered, not least through its passports that are 'valid as long as the bearer can still laugh', a sense of belonging and at-homeness in a state that many at the time perceived as cold, hostile and teetering on the brink of the next world war.

Contrasting visions of Europe

From a 'Fortress Europe' envisaged to keep the barbarians of all perceptions firmly outside the gates to a lost *Mitteleuropa* imagined as a peaceful melting pot of cultures with a Jewish intellectual class as its cosmopolitan heart, visions of Europe are numerous and very often conflicting. Except for the lost *Mitteleuropa*, few of these visions acknowledge 'those ecological polyphonies – material and imaginal – that found our common *oikos* ... our ultimate if always uncertain sense of being at home' (Biggs n.d.: 18). And while he has a lot to say about issues of dwelling and belonging, the loner from Todtnauberg tends to be rather over-cited and under-stood in these debates. But that is a matter for another day.

According to Donald Rumsfeld and others, there is a 'new' Europe and an 'old' Europe – the former ready to buy into America's mission to save the world from itself, the latter with a mind of its own. It was clear where his preferences lay, and they are not mine. One of the great issues of our time is the displacement of old Europe – ideologically, politically and even economically as the American free market does its best to pull the world with it into free fall. In this situation, we would do well to realise the subversive potential of being 'from here'. This relates to the existential ecological groundedness referred to earlier. Both historical and contemporary precedence can be found in the debatable lands that I have traversed here.

In Sarmatia, this is known by the term *tutejsi* (spelt or transliterated – from the Cyrillic – differently in different regions).[6] The category first appeared as *tutejszy* in connection with the Polish census (Trepte 2004). In 1931, only two 'ethnic' categories were allowed: religious affiliation (*wyznanie religijne*) and mother tongue (*język ojczysty*). This was to avoid problems and disputes that had arisen after the 1921 census when, in response to the nationality question, many non-Poles had been counted as 'Polish' because that was the language they used most regularly. However, a significant majority in the north-eastern region of Polesia had responded to the language question by saying 'own language' (*swój język*), 'our language' (*po naszemu*) or 'local' (*tutejszy*). Those who spoke 'local' comprised some 62.4 per cent of the inhabitants of Polesia; in official documents and statistics, this group was henceforth described as 'Polesians' (*Poleszucy*).

At the other end of Sarmatia, in Western Ukraine, the concept has also recently made an appearance. Here it has been explicitly linked with a 'very old', 'quasi-pagan' Europe (Ivakhiv 2006). Iain Biggs (n.d.: 28) observed that '[d]welling requires homelessness'. What those who are 'from here' in this sense are saying about their identity is not that each individual human being has many identities (and *Heimat* places) but that only the multiple forms of expression and definition – understood here in the literal sense of being *de finis*, 'about boundaries' – *taken together* make up the respective identity/*Heimat*. Where a number of people relate one (identity) or the other (*Heimat*) or both to the same bioregion, there we may have something like community in the ecological sense (see Deffenbaugh 2006).

In the city of Gdańsk, I hear a German man talking about *Heimat* to his granddaughter. His grandfather, whom he never knew, had died as a refugee crossing the frozen Vistula Lagoon in early 1945. He has been visiting the area occasionally since the end of the Cold War and now he is bringing his granddaughter to experience what he refers to as his *Heimat Europa*, which he has regained since 1989. It would be easy to interpret this as pure nostalgia. At least this man has some family connection with the region, unlike me, and therefore some basis for romanticising it. But that interpretation falls well short of coming to grips with what is going on here. He was born far away from the region and was only able to visit it as a middle-aged man. His daughter is already two generations removed from the area but she still sees it as important that her daughter should get to know the region, even as she fully realises that this is no longer the region where that daughter's great-great-grandfather once lived. Why? And what does the man in the middle of this five generation trajectory mean when he speaks of a *Heimat Europa* – this area, or something larger, wider, of which it forms an essential part? Does 'from-here-ness' remain a matter of hostile differentiation by language (Miłosz 2002: 29), or has it again, as Ivakhiv (2006) hints, become something more comprehensive and ecologically grounded?

Oral tradition in cases where the process of tradition happens away from the ecological context it refers to still has relevance for belonging and the imaginal components of our identities. But this obviously is not, cannot be, the same relevance as for people coming from a place. Coming home to a place that has not been our own is possible, but what does it mean for our reading of the presuppositions underpinning cultural expressions, old or new? Is there a way of connecting with an old Europe? If I say that I know there is, I don't mean 'know' in the sense in which we normally use that word. Perhaps a better answer may be found in the praxis by which we make our places habitable. But who do we have in mind when we do that? Is not all inclusion, whatever its motivation, also a form of exclusion by default? We may well need to develop some radically new categories and a corresponding analytical vocabulary before we can expect deeper insights into these issues.

Cultivating a field of place wisdom[7]

Around 1970, Joseph Kosuth (1991: 117, 119) issued this challenge that characterised the artist as role model for an engaged anthropologist:

> Because the anthropologist is outside of the culture which he studies he is not part of the community. ... He is not part of the social matrix. Whereas the artist, as anthropologist, is operating within the same socio-cultural context from which he evolved. He is totally immersed, and has a social impact. His activities embody the culture. Now one

might ask, why not have the anthropologist ... 'anthropologize' his own society? Precisely because he is an anthropologist. Anthropology ... is a science ... [and] ... is dis-engaged. Thus it is the nature of anthropology that makes anthropologizing one's own society difficult and probably impossible.

Informed as it was by an image of anthropology that held largely true before the beginning of that discipline's bumpy homecoming, the challenge may appear somewhat dated after some four decades of critical encounters between art and anthropology (e.g. Schneider and Wright 2005; Svašek 2007). Not only has there been increasing interest in the anthropology of art (Gell 1998; Morphy and Perkins 2005) an 'ethnographic turn' in art (Coles 2001) has also long been diagnosed, and anthropologists whose practice has been shaped by these and related debates (e.g. Greverus 2005) have long since 'anthropologised' their own societies.

In the following pages, no attempt will be made to review the extensive and growing literature dealing with the relationship between art and anthropology. Rather more humbly, setting out from a personal point of departure, I want to offer some reflexive observations and speculative thoughts on a very specific aspect of practice where art and anthropology may be seen to converge, pondering the creative potential of such convergence. The vanishing point towards which such a meditation is directed is, as it was for Kosuth, the social impact such practice might have. Within anthropology, the social side effects of its canonical disciplinary practices in the context of imperialism, colonialism and other forms of oppression have come under scrutiny for some time; but here I understand 'social impact' somewhat differently, in the sense in which the term is used in what is often called 'applied anthropology' or 'public ethnology' – approaches committed to emancipation through political intervention based on scientific analysis, and thus not a million miles removed from the founding ideals of a certain 'Free International University'. To frame this meditation, I will look at the 'site' of our practice, or what anthropologists have tended to refer to as 'the field', drawing loosely – some might say frivolously – on some rather heretical ideas. Kosuth spoke of anthropology as a science, by which he probably meant that it was committed to a certain epistemological and ontological paradigm. My starting point is a hypothesis (Sheldrake 1981) that I encountered as a young graduate reading up for a Ph.D. proposal while teaching German Studies in a school of accounting and applied economics – a hypothesis the publication of which prompted Sir John Maddox (1981: 245–6), then a senior editor of the journal *Nature*, to suggest the book might be the best candidate for burning in a long time.

Morphogenesis

The hypothesis in question was Rupert Sheldrake's 'morphogenetic field'. Literally, the term 'morphogenesis', which has been used in biology for some time, refers to the creation or coming into being (*genesis*) of a form or shape (*morphê*); Sheldrake uses it to express his idea of a living, developing universe with its own inherent memory. He argues, basically, that natural systems, by what he calls 'morphic resonance', inherit some kind of collective memory of patterns of behaviour and even physical development. Sheldrake proposes a continuous spectrum of such 'morphic fields' including morphogenetic, behavioural, mental, social and cultural fields. Existing within and around the system that they organise, they contain the collective memory members of a species may draw on and to which they contribute. Thus the fields themselves evolve. According to Sheldrake, 'morphic fields' contain information but neither matter nor energy, and they are discoverable only through the effects they have on the systems that they are part of. Social fields, for example, influence the behavioural patterns of all individuals constituting a particular social system; cultural fields shape how traditions are transmitted across space and time.

As an explanation of the modes of transmission of concepts and archetypes, Sheldrake's theory appears to owe much to C. G. Jung's (1991) theory of the collective unconscious, which Jung saw as a deeper, in some sense ultimate, biological reality. The hypothesis of a universal field that encodes 'basic patterns' can be traced back to the scientific world view developed by Johann Wolfgang von Goethe (Bortoft 1996); with Goethe, Sheldrake also shares the notion that this field holds the information needed to bring forth animate beings and their behaviour and to coordinate patterns of existence with those of other such beings. In other words, the field provides a force guiding the development and growth of organisms so that they take a shape similar to others of the same species – and this applies both in a physical and in a sociocultural sense.

The biological significance (or otherwise) of Sheldrake's theory is of minor concern here but his ideas deserve consideration with regard to the social and cultural fields as such, and also in another, related sense, which shall be the main focus here. As ethnologists and anthropologists we are used to thinking about fieldwork in a relatively 'physical' sense of being – often for an extended period of time – at a particular 'site' where we study the lives of animate beings usually belonging to our own species. Kosuth thought this worked better if we did it in social and cultural fields other than our own; for a long time, and in many instances right up to the present, the anthropological establishment tended to agree.[8] However, if for the sake of intellectual curiosity we accepted Sheldrake's model, however contested it may be, we can see cultural and social fields as entirely metaphysical – not discrete physical locations that we enter and leave at will but webs of memory and belonging in which we may become entangled and which we may find difficult to disentangle ourselves from.[9] Undertaking fieldwork in such circumstances also entails the possibility – if not indeed the danger – that the researcher might alter the memories and relations that give shape to the field, becoming part of them, modifying in the process the subject of study in ways utterly anathema to the scientific paradigm.

Working the field

In anthropology, the paradigm of long-term 'residential' fieldwork is commonly traced to Bronisław Malinowski's prolonged involuntary sojourn on the Trobriand Islands, where he developed what later became a key anthropological method, called 'participant observation'. Long-term fieldwork was, however, not invented accidentally during the First World War, as this well known narrative might suggest; it has a somewhat longer pedigree. One of its chief proponents was the German scholar Wilhelm Heinrich Riehl, one of the ancestors of European ethnology, who in the second volume of his epic work *Land und Leute* (Land and Folk) characterised the task of the ethnographer thus:

> To roam freely through the world, the eyes always open for nature and the folk makes cheerful work – a jolly game it is not. ... In my view the double work-load of simultaneously roaming and investigating is especially strenuous, far more strenuous than the most thorough study of books at the desk.[10]

The emphasis for Riehl was very much on 'wandering' around the field with open eyes, ears and mind, the whole sensory apparatus fully alert; moreover, he saw direct engagement with the facts of life as the best source of ideas and the immediate recording of one's observations (as well as the ideas they give rise to) as a vital ethnographic practice.[11] Existentialist and phenomenological fieldwork that became fashionable in anthropology in the second half of the twentieth century looked in many ways not unlike the methodology advocated by this cultural historian a century earlier.

Over the past generation or so, anthropology has increasingly discovered 'home' as a theatre of fieldwork. The reasons for this 'homecoming' are manifold and cannot be discussed here, but the ambivalent notion of 'theatre' is worth highlighting. The term is used both in the sense of 'stage', designating a space for drama performances, and similar to its medical use (as in 'operating theatre'), indicating the performance of more or less complex operations. With the emergence of virtual 'cyber-reality' and other disembodied forms of being, such as financial futures markets where profits not yet made from the exploitation of resources not yet located are sold for profit, the question of what constitutes 'the field', and how and where it is situated in time and space, has become a matter of considerable debate. As the editors of a recent book on the topic have observed (Coleman and Collins 2006: 11), 'the anthropological tendency to argue and think through spatial metaphors has concealed the degree to which fieldwork has never been dependent on fixed places as such'. Arguably, the places we enter are, in some ways, invariably imagined places and, as such, not fixed in any reality outside of this. Or so the fashionable constructivist theory of the world would have it. The ontological argument over whether or not there is any reality outside our perception must be left for another occasion; from a practical perspective, I prefer to use the term 'actuality' (*Wirklichkeit*). The term refers to what effect the world 'outside' has on an individual or group, regardless of whether or not there is such a world in philosophical or even purely physical terms, and thus allows us to get on with trying to make sense of the world. Obviously, if this actuality has significant and predictable effects we may have some grounds for surmising that there might be a reality, and what it might look like.

Postmodernity is credited – or should that be: discredited – with the progressive destruction of place, and the anthropological departure from conventional notions of fixed places has been part of that process. However, seeing places as performances does have analytical advantages when we are trying to come to grips with fieldwork. Such a perspective 'captures a sense that fields (and associated relevant "contexts") are created anew each time the ethnographer ... invokes the field in the process of research and writing. ... [T]he field as event is constantly in a process of becoming, rather than being understood as fixed ... in space and time' (Coleman and Collins 2006: 12). And we do not have to subscribe to a constructivist view of the world to acknowledge that 'fields are as much "performed" as "discovered", framed by boundaries that shift according to the analytical and rhetorical preferences of the ethnographer' (op. cit.: 17).

As fieldwork has become increasingly detached from the conventional anthropological 'field' – 'a "tribe", a village, some place you could get to know by covering it on foot and engaging with its people face to face' (Hannerz 2006: 23) – we have discovered 'multi-sited', 'yo-yo' (Wulff 2007) and other strange and novel fieldwork styles. Nowadays we often seem to have little idea of 'what the field is, or where it should be, if it is real or perhaps virtual, and even if there has to be one at all' (Hannerz 2006: 23). As Cristina Sanchez noted at the Europeanist anthropology symposium mentioned earlier, ethnographic fieldwork is not 'qualitative' as opposed to 'quantitative' research – it is something different. Few outside the practice of fieldwork seem to understand that difference.

Initiated in the 1980s, the 'writing culture' (Clifford and Marcus 1986) debate has continued to challenge the ways anthropologists construct 'their' fields and the people they write about, but 20 years on it is by no means certain whether 'we have yet adequately faced up to the problematic of diversifying anthropological genres' (Hannerz 2006: 33). Indeed, we continue to write as if the community studies paradigm still prevailed and are only gradually coming to grips with the need to adapt how we write to what we write and what we write about – we might even want to interpret 'writing' rather more loosely than has been the case to date, and 'should be wary of allowing the routine assumptions from established styles of fieldwork to carry over into and dominate arguments about newly emergent styles' (Hannerz 2006: 33).

Because anthropology was conventionally focused on exotic locations and on peoples without 'written culture' in the way the dominant global cultures have understood that term, it neglected the historical dimension for a long time. As the discipline was 'coming home' it began to discover the past (including its own) as a suitable – and often suitably foreign – field. Moreover, as well as the pasts of the groups of people it studied, anthropology became interested in the pasts of individuals and so biography and autobiography came to be recognised as valid pursuits for its practitioners. The inclusion of biography and autobiography as spheres for research has had a significant impact on the definition of 'field' for the purposes of anthropological fieldwork because 'life is always lived in certain places that need to be investigated in the course of the research', and where the focus is on an individual, 'the life of the subject becomes the field constituted by the events of the life' (Kristmundsdottir 2006: 168). As indicated earlier, the actuality of such a 'life lived in places' does not, strictly speaking, require the physical reality of a place outside of that life as a necessary condition (however useful it might be for practical purposes). It is reasonable to postulate a place constituted by life events; but if we accept that postulate, we must also accept that the 'field' where ethnologists/anthropologists conduct their fieldwork can no longer be 'a place or locality in the traditional sense'. As the conventional boundaries dissolve, 'admitting new kinds of research and access to new kinds of knowledge' (Kristmundsdottir 2006: 169), we need to re-orientate ourselves. Because of its emphasis on participant *observation*, anthropology has often been criticised for being overly concerned with the spectacle, with what can be seen. Where key actors are dead, the researcher inevitably must 'use her eyes rather than her ears to discern the voices in the field'; nevertheless the past 'is a field like any other in anthropology, teeming with voices that speak to the anthropologist' who is able to listen (Kristmundsdottir 2006: 170). Following Sigridur Kristmundsdottir's approach, the participant observer may listen carefully to *voices in the field*, 'albeit subjectively, since in interpreting these voices a researcher has to rely on imagination to a greater extent than in the traditional field' (op. cit.: 175). Here, then, is another heresy.

Fieldwork has other, non-academic connotations: as the work of an agriculturist. Strictly speaking, a farmer does not grow a crop; crops grow by themselves. The farmer sows and plants and then works on the local microcosm to create the conditions conducive to the growth of those crops. Thus the farmer cultivates the field, creates a culture – hence we speak of an agri*culturist* – the fieldworker who sows and plants, then cultivates the field, influencing the patterns of plant growth in order to help a culture to emerge, take shape, drawing on the morphogenetic field within which all this happens.

When I teach ethnographic fieldwork methods to students, I like to recount an anecdote from the Zen tradition:

> A frazzled disciple, wilted by the noonday glare, goes to the master for advice. 'Sensei!' he pleads. 'Counsel me! I no longer have any hunger for enlightenment. Tell me what to do!' 'Have you a garden?' the master asks. 'Yes.' 'Then go hoe it!' 'But what about enlightenment?' says the puzzled disciple. *'Forget enlightenment!'* the Sensei roars. *'When you hoe – hoe!'*

> (Walters 2001: 46–7; original emphasis)

The point of telling the story in that context is to emphasise the tedium and drudgery of much of fieldwork where for most of the time very little happens that might be considered exciting – intellectually or in almost any other terms – and a fieldworker might well be tempted to pack up and go elsewhere, but should instead stick it out and do something in keeping with the needs of the local microcosm, even if this activity may yield any 'useful results' only in the fullness of time. But the image that this little gem of fieldwork wisdom invokes in my mind is neither of a diligent farmer nor of a wise master of Zen. It is of the artist Joseph Beuys (see Kockel 1995) planting an oak tree next to a column of basalt.

Beuys was not only an observer and listener but his actions, or at least some of them, can also be regarded as empirical research (Lerm Hayes 2006). *Empirische Kulturwissenschaft*[12] – empirical cultural studies – is one of the many names my academic discipline, European ethnology, is known by. Consider Beuys in the light of this meditation so far – a participant observer and listener, conducting empirical research on cultural issues, planting trees as he goes along. It would not be unreasonable to describe him as a fieldworker: anthropologist – agriculturist – artist. Thinking further along these lines may open up new genres, new ways of doing fieldwork and new ways of speaking about it.

There is also, not to be forgotten, the political aspect.[13] A widely quoted passage from Martin Buber's *I and Thou* (1970: 60–1) comes to mind:

> What is required is a deed that a man does with his whole being: if he commits it and speaks with his being the basic word to the form that appears, then the creative power is released and the work comes into being. ... Such work is creation. ... As I actualize, I uncover. I lead the form across – into the world of It. The created work is a thing among things and can be experienced and described as aggregate of qualities.

While he could be interpreted as talking about the artist working in a kind of social vacuum, 'Buber was first and foremost a prophet of community' (Deffenbaugh 2006: 146). In his framework, 'community' can be seen as 'a certain refuge for those plodding alone through the meaningless sands of the It-world. For Buber, therefore, the care of souls is inextricably linked to the care of soils' (loc. cit.). The appropriate context for this process of community is not the universe but 'a particular place whose peculiar creations – narratives, liturgies, music, rituals, art – are faithfully passed on from one generation to the next' (loc. cit.). Thus we are dealing with a process of tradition – which is a key concern in European ethnology.

Anthropos, Ethnos and Topos

To the extent that the social sciences have turned their searchlight increasingly towards such allegedly universal forces as the market or globalisation, they have become almost devoid of meaningful notions of the *anthropos* – their focus on metaphysical forces that bump a cultural construct around has eclipsed concerns with understanding human being. Postmodern analysis has debunked the quest for the *anthropos* by postulating the inevitable constructedness of the individual as a non-negotiably individual matter, thereby rendering any alternative approaches beyond the pale. Even some anthropologists have fallen into this discursive trap. By the same token, ethnology has been stripped of the *ethnos* as 'community' has been turned into a bland and meaningless concept, the term nowadays being used fairly indiscriminately to describe any analytical category of more or less faceless people for the purpose of social analysis. If 'community' can be defined in any way we like and individuals are individually constructed, we are confronted with a stark choice: we can either abandon anthropology and ethnology as pursuits no longer relevant in a postmodern nirvana or we can take a stance against this politically correct voiding of the *anthropos* and the *ethnos* that goes hand in hand with the discursive destruction of the place – the *topos* – they inhabit. To take that stance makes our fieldwork inevitably a political act of defiance, which may well lead us onto potentially dangerous ground, where we need to know what we are sowing and be prepared to deal with what we might reap. In other words, we need to take full responsibility, not just for and towards a scientific establishment but for and towards all human beings that we engage with. This responsibility is not simply something 'we' must face vis-à-vis 'them'. As fieldworkers we make the field, but the field also makes us. Nobody has expressed that better than the Dutch artist Maurits Cornelis Escher in his *Drawing Hands* (1948): the anthropologist writing about the *anthropos*, the ethnologist writing about the *ethnos*, is like *the* hand that draws *a* hand that draws *the* hand that draws *a* hand that draws ...

Many years ago, the magazine *The Ecologist* carried a cartoon showing a group of scientists in motion: one dreaming up a complex formula, another gazing at the sky through a telescope, a third studying a flower with a magnifying glass and their leader, his face buried in a bundle of papers marked 'data', pointing ahead cheerfully while striding forcefully forward, over the edge into the abyss. This image of a self-absorbed science is aligned with the Enlightenment and juxtaposed with an alternative view of knowledge and understanding that is aligned with the Romantic

Movement. The epistemological resonances of this juxtaposition cannot be explored here – nor can the validity of the juxtaposition be examined. But two observations may be noted that are of relevance to this meditation. One is that the constructivist paradigm fits in well with the kind of science caricatured, rather than with its alternative, as post-postmodern proponents of that paradigm tend to perceive; and this is because, second, it shares the same disregard for the place – the *topos* – in which it is situated.

Fieldworkers live and work in constant interaction with human beings in communities of place and may therefore be less inclined than desk researchers to deny their materiality. However, the political imperative indicated above requires that we do not stop there but develop, and engage with, a vision of place actualities that runs deeper than the mere acknowledgement of their existence. A century ago, the German Youth Movement used the metaphor of 'The Blue Flower' to give expression to such a vision;[14] Beuys may well have been influenced by their ideals, especially where they touched on and were inspired by anthroposophical concepts. Much of his work can be understood as engaging with questions of humanity, community and place, which he sought to both depict and shape – a fieldworker in the morphogenetic cultural field.[15]

Ethnologic

In a series of 'essays on cultural renewal', the poet Kenneth White (2004b: 22) postulates the urgent need for a new anthropology: 'The real work consists in changing the categories, grounding a new anthropology, moving towards a new experience of the earth and of life.' In pursuit of such a new anthropology, he invokes an aspect of the Romantic Movement that we have encountered earlier in the work of Wilhelm Heinrich Riehl (White 2004b: 96; original emphasis):

> Romanticism meant ... a radical crisis in the Western conception of the world, a criticism of its systems, values and ambitions, an encyclopedic search for knowledge in all directions and the groundwork for a new epistemology, as well as a tremendous outburst of creativity. A lot of this was expressed in ... the 'transcendental travelogue' ... [which] ... moves through a spiritual topography ... It is a journey from self to Self, from confusion and ignorance to a cosmo-poetic reading of the universe. But more important perhaps than the destination of these transcendental travelogues is their *method*. The idea is to give a sense all along the way of what is open and flowing and cannot be defined in any cut-and-dried fashion. ... All is essay, fragment, approach.

In other words, he postulates an interactive, creative approach to fieldwork taking *poesis* literally when he insists on the need for 'not only a new philosophy of poetry, but a new poetic anthropology' (White 2004b: 145). And here, again, we can see Escher's hand that draws *a* hand that draws *the* hand that draws *a* hand – a representation that reminds us also that, contrary to the universalist aspirations of paradigmatic science, 'all logics are ethnologics' (Kan and Strong 2006: xvi). They are grounded in place and community. Ethnologically speaking, the key universal is that there are few universals.

Another Scotsman, the Gaelic scholar Donald Meek (1995: 34), said of the aims of an earlier generation of Scottish poets that they 'focused in one word – "community"'. Yet another Scots poet, the human ecologist and activist Alastair McIntosh, quoting Meek, has argued that '[i]t is precisely this "Celtic" sense of community that the casualties of globalisation, which is to say many people in the modern world, turn to for a bit of vision, hope and nourishment' (McIntosh 2002: 19).[16] In that context, '"Celticity" ... takes on a meaning that can be bigger than ethnographic and linguistic definitions alone: it becomes a code for reconnection with human community, with the natural world, and with God.' It was precisely in this sense that I interpreted the 'Celtic Quest' of Joseph Beuys (Kockel 1995). The theological resonances of this position cannot be explored here; but what becomes apparent is the triad that points beyond the traditional confines of anthropology as perceived by Kosuth and other critics: the *anthropos* reconnects with the *ethnos* in the *topos* – or in Geddes' terms, work-folk-place (see, e.g. Law 2005). We could extend the line by saying that this process creates the *oikos*.

Understanding anthropology and ethnology in these terms, and conceiving of their 'field' accordingly, leads to a fundamental reorientation of practice for epistemological and ethical reasons. Many anthropologists and ethnologists may already see themselves in a similar light, and often their choosing this discipline for their career has been motivated by concerns not unlike those that have inspired the work of artists, poets, theologians and campaigners. Hence there should be, and is, considerable potential for – to quote Geddes again – sympathy, synthesis and synergy (see Stephen 2007: 23). But in practice we are all too often afraid to make the connection. The development of what might be called a 'deep anthropology/ethnology' has been hampered by the contemporary fashion for cultural studies that focus on the 'booming, buzzing confusion' we encounter at the surface of the everyday, and by an ever tighter funding regime that values short-term instrumental usefulness that can be readily proven rather than long-term elementary understanding that can only be conjectured in the present. And so the fieldworker, just like Kenneth White's modern citizen, '[h]ustled hither and thither between bureaucracies and circuses, between boredom and distraction,' is no longer able to find his or her bearings 'in a civilization which, having no deep culture, tries to camouflage its fundamental lack by making a lot of noise and flashing a lot of images' and therefore 'ends up avoiding any still, open moment ... and, more or less satisfied, but knowing little real joyance, lives on complacently in a well-filled mediocrity' (White 2004b: 59). This complacent mediocrity is the context for and the reason why a reorientation is arguably necessary, for epistemological and ethical reasons, unless we are prepared to abandon the field – thereby literally losing the plot – to the property developers and financial speculators. As ecologically aware fieldworkers, we are called to engage more actively with other forms of understanding, such as can be found in art when it is 'founded and grounded, that is, when it isn't just another aspect of the circus' (loc. cit.).

Is a particular ethnologic implied by the citation of Scottish authors in this section?

The wandering trickster

Cultivating sympathy, synthesis and synergy between ethnological/anthropological fieldwork and art at multiple levels of engagement is vital for the continued meaningfulness of anthropology and ethnology as intellectual pursuits. This may seem a bold statement. We are faced with a strong paradigm that has almost succeeded in digging the ground from under our feet by declaring as void – at least in academic terms – any attempt to connect with deeper levels of meaning. The charge levelled against any dissident is that of 'essentialism'. Yet as I have argued in Chapter 5, labelling certain ideas and avenues of inquiry as 'essentialist' is a convenient way of avoiding issues that might challenge fashionable normative frameworks. There is a danger that we fail to recognise the myth of 'essentialism' for what it is; and we may even forget how important it is 'to keep the power of myth under a watchful eye and to offer narratives that challenge and critique the dominant worldview created by our stories' (Deffenbaugh 2006: 161). Cultures[17] across the world have used so-called trickster tales as a corrective to myths. Where myths are supposed to establish a sense of permanence, certainty and security with regard to cultural frameworks and practices, such tales serve as reminders to the members of a particular culture 'that the reality they perceive through their myths is nonetheless vulnerable to disruption, that disorder is as much a part of their waking day as the order they have come to know and expect' (Deffenbaugh 2006: 161–2). The best-known trickster figure is Coyote.

Coyote's primary purpose is to change things, enabling what is old and outdated to disappear and the world to be renewed. Consequently, in some creation myths Coyote becomes the Creator; in other instances, he may be a messenger, a culture hero or a fool. Coyote is able to shape-shift. His creative power extends to language – he can create animals by naming them. In such myths, Coyote is not referred to as an animal and may even meet his animal counterpart, the coyote.[18] In such encounters they usually address each other, respectively, as 'elder brother' and 'younger brother'. In many traditions, Coyote appears as a culture hero engaged in changing the course of rivers or the location of mountains, creating entire landscapes or obtaining sacred items for humans.

Coyote – the artist, poet, wandering fieldworker in the morphogenetic cultural field? If this is a general, 'global' image, is the Scottishness of earlier references then purely incidental? Well, not quite. Coyote has also been observed on Rannoch Moor, that desolate place in Highland Scotland that inspired Joseph Beuys to his work *Celtic (Kinloch Rannoch) Scottish Symphony*.

Roaming in upside down places

We have finally arrived at the centre of the labyrinth, in a place that, in its bleak desolation, perhaps epitomises the 'Celtic Fringe' better than any other. A poster by Scottish Heritage – on display, for example, at the Heritage Centre in Glencoe, which commemorates the 1692 massacre of a branch of Clan Donald – depicts a fierce-looking Highlander posing in a bleak heather landscape, with a caption that reads: 'Once they owned it. Now, it's yours.' While the caption is probably just meant to convey the message that national heritage is now accessible to all, the poster gives the visitor no sense of the individual and communal pain endured by the commoners who were displaced during the Highland Clearances; it thus gravely misrepresents the story of the place in which it is now being displayed. As we move out of Glencoe we come to Rannoch Moor.

The significance of this landscape for Beuys and his work in relation to Scotland and Ireland has been explored in depth elsewhere (Walters 2009; also Rainbird 2005). What is important to highlight here is the obvious point that, as an artist, Beuys was not making an ethnographic statement about Rannoch Moor in the way that a diligent ethnographer would make such a statement. At the same time, the universal artistic statement was framed unmistakably by the experience of a particular place. Unlike universal statements based on the analysis of so-called hard data, this universal statement retains resonances of its local roots and dispositions that make it relevant and applicable to the local context. It is this curious specificity of the universal that I see in the *soziale Plastik* postulated by Joseph Beuys, as I see it in the 'geopoetics' of Kenneth White. Both treat the concept of 'tradition' quite differently from the way in which we have been used to thinking about them – as a force of considerable creative potential, rather than an obstacle to cultural advancement. Naming his *Scottish Symphony* after a place on Rannoch Moor, Beuys – perhaps inadvertently – gave his audience a hint in that regard. 'Kinloch' means 'the Head of the Lake', but the place itself is actually at the opposite end to the head of Loch Rannoch – an upside down place, so to speak.[19]

The heritage and traditions of the displaced Highlanders may well contain elements that from our contemporary perspective seem hardly worth having; but to jettison an entire culture because it appears to have some room for improvement is a rather drastic solution. Nothing can justify ethnocide. Rather, 'the soils of our tradition need to be turned, and new seeds need to be sown, but from stock that we have inherited from our predecessors' (Deffenbaugh 2006: 214). If we abandon the quest for humanity (*anthropos*) constituted through community (*ethnos*) in its place (*topos*), in favour of more fashionable or temporarily lucrative knowledge interests, we risk losing the ability to tell good seeds from bad ones, and we might even be tempted to regard 'our' seeds as better than everyone else's. It is important that '[w]e do not seek to "set upon" our place in order to fulfill our own needs and desires. Rather, we endeavour to be that point in creation at which art and local ecology are joined in unison so that the Eternal You may come to presence as a special kind of language' (Deffenbaugh 2006: 194; see also Spirn 1998).

Toposophy

Are artists really better anthropologisers of their own society than anthropologists, as Kosuth once claimed? Perhaps some are. More important than whether artists should turn themselves into anthropologists or vice versa is the question of what we might be able to learn by mutual exchanges, not just of ideas but of actual practices. As anthropologists are coming to terms with the fact that they make, and are made by, the field that they study, they have a choice – retreat into the safe realm of pure cultural theory or get to grips with the messy business of trying to navigate the morphogenetic cultural field as it changes shape under their very hands. Does that make them social sculptors? If so, it is important that they understand fully where they are and who they are with when they do this. Ecologically, human beings live in communities in specific places, where roots are put down and pulled up in the course of time, and memories make dreaming the future possible. Beuys the trickster, wandering around the 'radical field' (McManus 2007) on the 'Atlantic edge' (White 2006) like any old-fashioned ethnologist concerned with wisely applied politics, has demonstrated in his actions how the wisdom that 'sits in places' (Basso 1996) may be released. It is not important here whether he always succeeded in that, or whether his was the right – or even the best – way of doing it. What matters is that he tried, and tried to show others how. Anthropologists could do a lot worse than follow him 'across the territories' (White 2004a) of perception in search of enlightenment. But we shouldn't forget to bring our hoe ...

Concluding thoughts

Throughout this book I have raised questions rather than offering definitive answers. I set out to look for Europe. This I have done in an oblique way, rather than explicitly and analytically, by rummaging in various places. Where is Europe? Can we find it in the frontier as understood here? I am tempted to say: only in the ethnic frontier can Europe be found. Neither the ethnically homogeneous Europe of certain dictators who very nearly destroyed it nor the blissfully indifferent Europe populated by narcissistic individuals that is favoured by the free-marketeers and their postmodernist accessories; but the Europe of cultural encounters that are structured by interweaving trajectories. To develop a vision of a truly 'new' Europe, we do need to track these trajectories, and in that sense a focus on flows (Augé and Colleyn 2006) is indeed important. But in our concentrated efforts to discern the forces and directions of global flows, we do need to keep an eye on the actors without whom nothing would flow. Every so often, we hear that interest rates have risen. Interest rates do not rise – they are being put up, by committees comprising human actors making decisions. Likewise, goods, money and ideas do not flow of their own accord. People at least may do so. That is a crucial difference. In moving, staying, and mixing, people create memories, which in turn are the foundation for trajectories and invest them with historical legitimacy and credibility. As ethnologists/anthropologists, we observe these patterns and processes. Whether as human beings at home in the frontier we should also, like Kosuth's artist, take a hand in shaping them or whether that conjunctive implies an ethical imperative, and, if so, what form our interventions might take, are questions for another occasion.

7
Envisioning the Not-Yet: *Before Coming Home*

At the end of our travels, we may return home. Deepest Franconia, Lutheran heartland: a smallholding basket maker's home in the 1930s. The Sunday treat for the family of eight: one cured herring bought from a travelling fishmonger and laid out in the centre of the table so that potatoes could be wiped on its skin, for added flavour. We do not need to go back there (or pretend that we could) and we certainly do not need to romanticise it. But we do need to know about – need to understand it, because this is what made us who we are – the everyday lived experience of the generation that reared us. Would I be here without that herring (not a red one, I should add)?

There is much emphasis nowadays on applied, policy-oriented and vocational study, focused squarely on the contemporary. However, by wrenching the contemporary from its historical foundations, we are destroying one of the key functions of the contemporary itself – to become the past and the roots of our future. In the age of virtual history, we do well to remind ourselves that not every past is consistent with every present.

Heimat and heritage

Following on from my earlier explorations of the ethnic frontiers of European integration (Kockel 1999a), I suggested at the 2004 conference of the European Association of Social Anthropologists in Vienna that one possible way of approaching questions of European identities and boundaries may be to examine whether, and to what extent, Europe might be considered as being built by people who dwell in the frontier created by EU integration. In academic literature as well as in EU policy discourse, the term 'frontier' is commonly used as synonymous to and

189

interchangeable with boundary and border. In this confused, sloppy usage, the analytical potential of language is sacrificed. Etymologically, 'frontier' can be related to 'in front of', that which lies ahead, between the subject and his or her horizon. Therefore everywhere outside the 'self', and as far as the individual world version stretches, is the 'frontier', which thus forms an essential part of self-identification. Whereas territory is about spatial identity, and territorial space is defined by boundaries, the frontier can be seen as an indeterminate, open space. I have introduced the German term *Fremde* as translation of 'frontier' as a means of analysing the relationship between the 'self' and the 'other'. The frontier in this sense is outside one's own personal space, but nevertheless located within a familiar horizon – otherwise it could not be recognised as such, and could not serve as a parameter of belonging.

When I speak about Europe being built by people dwelling in the frontier, the terms 'building' and 'dwelling' have obvious resonances. Heidegger's philosophical wordplay with the terms 'constructing' and 'cultivating' – the German word *bauen* can mean both – suggests a link of the cultural with the political that is pragmatic rather than metaphysical. At the same time, however, the particular link implied, between place, being and longing, points, via an English wordplay in a Heideggerian spirit, towards the metaphysics of belonging – to 'be longing' as a general human condition – that can be expressed in the act of dwelling (*wohnen*) as inhabiting (*bewohnen*) a place or space in a way that invests the location with shared meanings, which over time become the habits (*Gewohnheiten*) of its inhabitants (*Bewohner*). These habits, as traditions and heritage, are then available as both repertoire and backdrop for autological and heterological representations of identity (Chapter 5). The latter have the 'other' as their 'target audience' and project a public image whereas the former are directed towards the 'self' and serve to affirm one's roots, imagined or otherwise. Elsewhere I have developed the contrast of territory and trajectory (Kockel 1999a: 290ff.). 'Territory' hardly needs clarification – its meaning as 'bounded area ruled by a particular power' is fairly widely employed. 'Trajectory' has its root in the Latin verb *traicere*, going across, and tends to be used to describe transgression, transition, or carrying across a distance, usually incorporating the overcoming of an obstacle. I have been using the term to denote the 'life-paths' (of individuals and groups) that link up significant places and/or events (Kockel 2005).

It is worth emphasising the implications of this use of terms. The character of the frontier, its *Fremdheit*, is determined – in the dual sense of that term – within the self. If something or someone is experienced as *befremdlich* (strange), that is not the same as *unbekannt* (unknown); rather, it means that the person or object has already entered the familiar horizon and is within sight (and potentially, one may speculatively pun, with insight) of the self, forming part of its world. Ontologically, the inevitable conclusion is that the self is at the centre of the frontier (*das Zentrum der Fremde ist das Eigene*) and that the frontier emanates from the self into the world (*geht vom Eigenen aus in die Welt hinein*). Thus by dwelling in the frontier and cultivating it, the inhabitants make it habitable and thereby gradually convert it into their own. This is the untranslatable, utopian place called *Heimat*, which is conditional on the *Fremde* as the horizon within which it can be experienced. Without the *Fremde*, as at once its frame of reference and its counterpoint, the idea of *Heimat* makes little sense. Moreover, the frontier serves as a kind of 'ante-room' for the 'self' – an 'other' may be safely translated into the familiar frontier and thus brought closer without actually being granted access into one's *Heimat*, thereby remaining *fremd* until he or she ceases to be *unheimlich* and perhaps even becomes *heimisch* (settled). If there is a *Heimat Europa* out there somewhere, it makes sense to look for it, and for the Europeans who make it, in the frontier – in open, undefined spaces and along trajectories.

The foundations of the perspective I have been trying to develop here reach back, as I have traced in Chapter 4, to a very different topic, namely the contribution of informal economy to regional development. A first tentative formulation of this perspective was published in a book-length study of three European border regions – Ulster, Euskadi and Schleswig – contrasted with contemporary migration (Kockel 1999a). Since then I have undertaken short periods of fieldwork in a number of other European border regions, including Lithuania Minor, but primarily among migrants, in particular Germans and Central/Eastern Europeans in the British Isles (Chapter 3). Alongside this fieldwork, I have followed up clues from an earlier episode in the ethnographic field that have led me into the terrain of anthropology and art via the historical and philosophical study of counter-cultural movements.

The ideological, or if you prefer the more neutral term, theoretical context for my present reflections on this research is rooted in deep ecology, an approach that emerged, along with my generation's passage into adulthood, during the 1970s. The intellectual and historico-cultural sources of deep ecology (Devall 1980) include Eastern spiritual traditions as reinterpreted in relation to modern physics (Capra 1975); minority traditions in Western thought; and a relational total-field image of human situationality, derived from ecological field theory, and in sharp contrast to the conventional view of man-in-environment. Arne Naess (1973), one of the key thinkers in the field, defined the central intuition of deep ecology in terms of biocentric equality (ascribing intrinsic value to every being, animate and inanimate alike) and 'Self-realisation' – not in the sense of the narrow definition of the self as an individual Ego but rather meaning the sum of our material and immaterial identifications, reaching from our immediate social context into the world, and eventually involving definition of our identity in relation to the Whole, the All-is-One, the unity with Being in a Heideggerian sense.

At about the same time as 'deep' ecology was in the ascendant, anthropology – much maligned for its colonial past and increasingly underfunded for its overseas pursuits – began to 'come home' (see Jackson 1987). This coming home of the discipline, which coincided with my first forays into the field, gave rise to considerable introspection about the need to do anthropology differently, without generating conclusive answers to the vital question – how? Even without any rigorous statistical data to prove the point, it is probably fair to say that young scholars engaging with anthropology for the first time during that period came out with a view of the discipline considerably different from previous generations. This may have applied more to those, like myself, who entered it 'sideways', that is, from another disciplinary background. It was, of course, also the time when dusty, stuffy old European ethnology began to discover cultural anthropology. In that situation, critical observers from outside the immediate discipline challenged the new project. One of these was Joseph Kosuth (1991: 117ff.) with his proposal of the artist as a model of the anthropologist engaged. And yet those anthropologists who were 'coming home' became, perhaps not surprisingly, particularly concerned with questions of identity, belonging and what I refer to here as *Heimat*, in other words, with the very issues they were, according to Kosuth, ill equipped to grapple with.

In German-speaking Europe, and to some extent beyond, the matter of *Heimat* has experienced a discursive boom since at least the 1980s. Despite negative associations, not least with the kitsch cinema and folksy pop music of the post-war economic miracle, the term has always had currency among the undogmatic Left, where Ernst Bloch in the 1950s had postulated it as the key utopian location (Bloch 1978). Regionalist movements and, from the 1970s onwards, environmentalists appropriated the concept as one imbued with positive values. As I have suggested in Chapter 4, this concept can be analytically useful when juxtaposed to *Herrschaft* as a subversive principle. At the broader societal level, the *Heimat* debate gathered momentum with the television screening of Edgar Reitz's eponymous family epic in the late 1980s. This coincided with the entirely unplanned end of the Communist regime in the German Democratic Republic and the subsequent, equally unplanned unification of Germany, which gave the debate a new direction. Since then, the debate has continued with a stream of articles, pamphlets and books (e.g. Krockow 1992; B. Schmidt 1994; Hecht 2000; Schlink 2000; Türcke 2006).

The English-speaking world simultaneously experienced a similar debate, although its focus was somewhat different. That debate centred on the matter of heritage and its role in contemporary society. Like the *Heimat* debate, it has had links with regionalism, and it also has had its subversive fringe. But by and large it has had two more pragmatic foci, perhaps reflecting the well-rehearsed (e.g. Lepenies 2006) contrasts between a civilised transatlantic world and a cultured but stuck-in-the-mud continental Europe. One of these foci, which overlap to some extent, has been on the commercial utilisation of heritage and all the concomitant problems; the other has been on issues of authenticity, legitimacy and the ownership of heritage (Nic Craith and Kockel 2007). Like the *Heimat* debate, the heritage debate has been fuelled by and has revolved around issues of belonging to and alienation from regionally grounded (in a semi-literal sense) culture contexts. The materiality of these contexts has been hotly contested, not least in the course of the 'invention of tradition' debate that has led many protagonists to proclaim the inauthenticity of traditions on the basis that they are invented.

Despite the logical fallacy of such claims, constructivist interpretations of the world are enjoying considerable popularity in academic discourse and beyond. This is not the place to engage in epistemological and ontological analysis, but it should be noted that it is irrelevant whether heritage and tradition are invented or communities imagined – if they have actuality, that is, if they shape the immediate and wider habitat, their reality is generated in the process and it therefore does not have to be a *factum a priori* in order to be valid. If we consider as 'traditional' and 'genuine' only those ideas and practices that were concocted by only God knows who, and at a time well beyond the horizon of memory (and preferably of written history), then we commit the fallacy of a shallow essentialism equating age, measured in geological proportions, with veracity and legitimacy and denying these qualities to any creative act in more recent times. We need better criteria for establishing what is or is not an authentic tradition. 'Authentic' does not mean 'exclusive' or 'exclusionist'.[1] 'Traditional' does not mean 'good and valuable by definition'; not all traditions are worth having, but the distinction is a matter of ethical judgement rather than ontological certainty. Moreover, the historical dimension of belonging that is highlighted by the concept of tradition must be ecologically embedded; our past is always a past in its place, the place we come from – even when 'coming from' already implies that we are no longer there.

When Hermann Bausinger, writing about nineteenth-century Germany, spoke of a *freischwebende* (free-floating) *Heimat*, he made a similar point. Paintings of roaring stags in the forest wilderness and other elements of a popular *Heimat* discourse in that period detached that very discourse from any material base. It was nevertheless effective. A century later, the same principle, in the guise of what one might describe as a free-floating regionalism, was discovered by, among others, Basque militants who proclaimed a global Basqueness that had no material basis other than the ideological identification with the cause of Basque self-determination. As the project of European integration progresses, creating an ever more mobile population of Europeans as a fundamental precondition for the smooth functioning of a single European labour market, the ecological groundedness of belonging will increasingly depend on its actualisation through such free-floating associations. However disconcerting Kosuth's reservations with regard to the feasibility of an 'anthropology at home' may have been when he first expressed them, the detachment, postmodern or otherwise, of belonging and identity from their concrete regional foundations poses an existential challenge for anthropology in general. There is a danger that the subject will disappear in a nebulous discourse of construction, invention and imagination that celebrates the alleged inauthenticity of everything cultural and the consequent non-existence of any material basis for cultural expressions, including the rug under its own feet.

On our journeys, we have balanced on borderlines listening to voices from the past in Ulster and spoken to the mobile Europeans in Germany, Lithuania and the British Isles. In the marketplace where our futures are traded (or traded in – for what precisely is hard to fathom) we have wandered from West to East and back again, glancing at the economy on and from both sides. Following the example of the post-Enlightened Romantics, we took a grand tour to remember European heritage and tradition (although ours did not proceed on well-trodden paths in the Mediterranean sun, staying in harsher climates instead). And towards the end we joined the Magister Ludi, Joseph Knecht, going for a swim in a cold mountain lake, as did young Tito, who survived the experience, and the reader may be left wondering whether the boy went on to find the Blue Flower, his true vocation to humanity, and managed to re-place his Europe. Perhaps we will meet him again on another European *Spurensuche*, a mythical remembering of tracks and non-places. But that is for another day, and perhaps another story.

In the meantime, if we believe the Eurobashers with their hidden agenda, we may think we know all about bad old Europe, and that we have had enough of it. Globalisation will surely finish it off anyway and it is time for us to forget about it, to go beyond it. *Au contraire!* We have hardly caught a glimpse of everyday Europe. The non-place is a *terra incognita* – neither a lost paradise nor a bad dream to be shaken off on the shrink's couch, but a new-found land with much to explore. What we see will be a matter of historical and contextual awareness. We may have exciting stories to tell from our different perspectives. What we tell and how we say it is a matter of political responsibility. My search for Europe over the past few decades has seen me outgrow a youthful international-ism that knows no ties to local or historical roots, towards a perspective of what I might call an 'enlightened' localism. In the postmodern spirit that has dominated much of academic discourse over the last generation or so, roots in locality and story needed to be destroyed, or disparaged at least, to facilitate the unbridled expansion of neoliberalist capitalism, horizontally across the world and vertically into our minds and souls. Individual and communal ties, both with one another and with ecologi-cal communities of place, are detrimental to that expansion. We need to ask – whose interests are being served by the attempted dissolution of places and pasts? Who profits from selling us the constructivist identikit in the identity warehouse that will ultimately make us indifferent?

The progressive excision of 'culture' from the ethno-anthropological agenda in favour of the more civilised term 'society' echoes the dissolu-tion of places and pasts, and the contempt for 'old' Europe expressed by many whose political background and agenda we would do well to remind ourselves of. Anthropology as culture critique does not neces-sarily mean having to run down culture at every opportunity; a less destructive way of critiquing culture would be by grounding our analy-sis once again in actual fieldwork conducted in regions understood as meaningful ecological contexts, such as the 'little Europes' that Schlögel (2002: 64) talks about – locational spaces with boundaries that are somewhat blurred rather than neat and clear-cut. Much of our recent conceptual problems with 'regional cultures' may reflect our failure to appreciate this quality of ecological regions, which makes attempts to squeeze cultural expressions into singular categorical boxes difficult, if not entirely futile. When we find a way out of this dilemma and learn to appreciate anew the richness of boundaries and frontiers, we may at last be able to reclaim the debatable lands of a dis-placed *Heimat* Europe by re-placing them. Boundaries and frontiers entail a duty to take responsibility for one's own house, and opportunities to be a guest in one another's – landscapes of transition where we can belong even if we do not speak the local language (Schlögel 2002: 193). Balancing on border-lines among mobile Europeans trading our futures, we should remember to re-place Europe before coming home. To achieve this may well require recovering the power of definition from the transatlantic *Herrschaft* and, as winter turns into spring, involve a certain re-orientation – a re-Easting – of what it means to be a European.

Notes

1 Setting Out: *Europe – A Winter's Tale …*

1. For a longer discussion of the location of Europe from various perspectives, see Kockel (1999a).
2. The image of the dotted line on the map is used by several contemporary authors, for example Wackwitz (2005).
3. Margarethe von Trotta's *Die Bleierne Zeit* (The Leaden Time; Bioskop, 1981), a German film about the lives of two sisters, was loosely based on the biography of Gudrun Ensslin, a member of the Red Army Faction. The title, taken from a poem by Hölderlin, was meant to hint at the societal situation of West Germany in the 1950s and 1960s but has since become a phrase characterising the period when the Red Army Faction and similar groups posed a serious challenge to the still young West German state.
4. The party lost its seats in 1984, returned to the European Parliament in 1989 with four deputies, but failed to win seats in 1994 and 1999, when it won less than half the number of votes it secured in 1979. Not until 2004 was it able to repeat – and indeed exceed, in terms of seats if not votes – its first European election success. In 2009 it almost doubled its votes and seats at the expense of the conservative CDU. See http://www.wahlen-in-deutschland.de/beBund. htm (accessed 1 December 2009).
5. The title translates as 'Someone will win', or 'There will be one winner'. The minister for School and Further Education in North Rhine-Westphalia, Barbara Sommer, reminisced in a newsletter issued in connection with Germany's EU presidency in 2007 that *Einer wird gewinnen* 'focused on Europe and presented Europe as a fascinating cultural area with cross-border opportunities for personal development'. Like many in the first post-war generation, she thus 'gained the conviction that a united Europe reinforces and promotes cultural and economic interests'. (www.europa.nrw.de/newsletter/nlfebruar07_web_ de.html; accessed 7 February 2008; the name of the presenter is misspelled in the newsletter.)
6. The notion of 'European visions', 'visions of Europe', or indeed 'Eurovisions', has been popular for some time. Since my 2001 inaugural lecture at the University of the West of England, Bristol, which provided the impetus for this book, 'European visions' has served as label for a youth competition in Germany (Behnke 2006), a project of politico-cultural networking championed by the Goethe-Institut and culminating in a conference at Berlin's Kronprinzenpalais in 2004 (www.kulturbetrieb.com/gkmain_de/projekte/ eurovisionen.html, accessed 1 December 2009), a disappointingly patchy survey of 'good Europeans' (Frevert 2003) and a 2004 series of short films titled 'Europäische Visionen', commissioned to mark the eastern enlargement of the EU, to name but a few examples.
7. The subtitle of Briedis' book refers to the fact that Lithuanians were until quite recently a minority among the population of Vilnius.

8. See Kockel (1999a: 52–60). The debate about Central Europe is going through cycles of vibrancy and dormancy. Since around the time of the 'Eastern enlargement' of the EU we seem to witness one of the former, in which German attempts at coming to terms with the rediscovery of lost territories are playing a noticeable part.

9. Founded in 2005, Europäischer Austausch gGmbH works with NGOs in Eastern Europe, especially in Belarus und Ukraine. The Susan Sontag quote is dated 1995 (www.european-exchange.org, accessed 29 January 2008).

10. The ritualistic destruction of objects considered a temptation to sin frequently accompanied religious sermons in central Italy during the fifteenth century. This practice became known as 'bonfires of the vanities'. It continues throughout the centuries in different forms.

11. *If on a Winter's Night a Traveller* is the title of a novel by Italo Calvino (1992).

12. Extract from the poem 'Deutschland: Ein Wintermärchen' by Heinrich Heine, reproduced in Lutz Görner ed. (n.d.), *Heinrich Heine (1797–1856). Ein Lesebuch für Demokraten, und solche, die es werden wollen* (no place given, published by the editor). While there are numerous other, more 'mainstream' editions from which this extract could be quoted, the text given here is accurate, and it makes sense to cite the source where I first read it, as an indication of the wider context of zeitgeist which influenced that reading.

2 First Journey – In the Frontier: *Balancing on Borderlines*

1. The discussion of Drumcree is based on a paper originally presented at a Jean Monnet Conference held by the Centre for European and International Law at the University of Liverpool on 4 July 1998. I am grateful for comments from Gerard Delanty and Nanette Neuwahl.

2. The discussion of the politicisation of the cultural landscape is based on a paper first presented at the 32nd congress of the Deutsche Gesellschaft für Volkskunde at Halle/Saale in September 1999. I am grateful to the participants for their comments, especially Henry Glassie for sharing some of his insights from his fieldwork in Ulster with me. My presentation included some 30 slides which cannot be reproduced here.

3. Brian Graham (1997: 192–212) suggests that discursively invented landscapes do not generate identities because they are contested. What he seems to mean here is a common identity of all parties to the conflict as a social utopia. If such a meta-identity does indeed exist in Northern Ireland, it has little relevance in everyday life, unlike the more immediate local and group identities.

4. Apart from the extensive cultural, geographical and ethnographic work of Estyn Evans in the 1930s, 1940s and 1950s, a particularly influential text has been by Heslinga (1962).

5. For a detailed exploration of the two types, see Kockel (1999a). Implicit in the distinction between 'here' and 'now' on the one hand and 'there' and 'then' on the other hand is the distinction between the *Diesseits* (this life) and the *Jenseits* (beyond), which can be understood in two ways – metaphysically, and also as the everyday experience of boundaries and frontiers (the clear line that defines the 'here' and the zone of contact that brings the 'there' closer).

6. For an example, see Adamson (1982). He has produced a range of publications, both books and essays, with a fairly consistent message throughout.
7. On this point, see for example the classic anthropological study by Rosemary Harris (1972).
8. There should be no need to point out that here, quite untheologically and yet somehow religiously determined, the spirit of modern rationality comes to the fore in one of its uglier guises. Unfortunately, the extensive literature on the relationship between Ireland and Britain all too often features the rather truncated interpretation of this process as the flowering of a specific, culture-immanent racism of which the English in particular are claimed to be singularly guilty. In my view, what needs to be addressed instead is a deeper problem with the spirit of modern rationality.
9. Henry Glassie drew my attention to this fine but important terminological distinction.
10. See Brett (1996: 117); John Bunyan's book appeared (in two parts, 1678 and 1684) at a time marked by intense religious conflict in England, and to this day constitutes a key text for the self-image and world view of various Protestant groups in Northern Ireland.
11. This section is based on fieldwork for a comparative study of identity processes in three medium-sized Irish cities: Cork, Galway and Derry/ Londonderry. Work in Derry/Londonderry started with occasional visits and interviews in 1989, followed by what may well be termed as long-term participant observation since 2001. Initial findings were presented at the 34th congress of the Deutsche Gesellschaft für Volkskunde at Berlin in 2003 (Kockel 2005a).
12. The designation of groups for the purpose of that great academic pastime, categorical distinction, is particularly fraught in Northern Ireland. While there is a high degree of congruence between unionist/British/Protestant identities on the one hand and nationalist/Irish/Catholic identities on the other hand, this is by no means a matter of straight equivalences. In the circumstances, rather than trying to capture the numerous possible combinations of political, ethnic, religious and other identity ascriptions, I have chosen to stick with local usage which aligns the 'camps' in the name debate along religious divisions.
13. In most cases the orange is replaced by a yellow referred to locally as 'gold', which may have one of two reasons – a rejection of peaceful relations between the 'British' orange and the 'Irish' green, implied in the design of the flag; or, more likely, a subconscious heraldic citation according to which the Irish flag combines the green of Republicanism with the white and gold of the Papacy. The combination green-white-gold was in use before the Irish Republic was founded and appears in some of the older Republican songs. In kerb painting, the use of yellow/gold is more common than in bunting. Locals I have spoken to over many years have invariably been surprised by this observation, and while some have ventured interesting speculative explanations, none has been convincing so far.
14. On 'Bloody Sunday', British soldiers opened fire on a civil rights demonstration, killing 14 people. The events have been the subject of a major public inquiry, the report on which was due some years ago but had not been published by the time this book was completed.

15. The Orange Order was named after William of Orange, whom the British parliament invited to succeed, as William III, the deposed James II. Against this background of ousting the legitimate heir to the throne, it is interesting that the members of the Order nowadays define themselves partly via their loyalty to the Crown (and, if considered necessary, against parliament).
16. Some observers speak etically of an 'invented' culture. For the time being, I prefer to go with the emic understanding, for reasons I have explained elsewhere (Kockel 2007c).

3   Second Journey – In the Diaspora: *Among Mobile Europeans*

1. Findings of this project have been presented at a number of conferences, including a meeting of the Commission for Intercultural Communication of the Deutsche Gesellschaft für Volkskunde in Munich in 2000 and a symposium of the Institute of Sorb Studies at Bautzen in 2005. For comments and constructive critique, I am grateful to Klaus and Juliane Roth and to Alois Moosmüller at the earlier meeting, and to Elka Tschernokoshewa and Konrad Köstlin at the later one.
2. This section originated as an invited paper for a seminar on contemporary diaspora experiences, held in Munich in 2002. I am indebted to Klaus Roth, Christoph Köck, Alois Moosmüller and Péter Niedermüller for the invitation and inspiring discussions during and after the seminar.
3. This workshop was part of the series of research seminars, funded by the Economic and Social Research Council, which I mentioned in the introductory chapter. Selected papers from the workshop have been published under the workshop title, *Communicating Cultures* (Kockel and Nic Craith 2004).
4. As this book went to print, Britain elected a Conservative-Liberal government. Policy directions will change as a consequence. Having been chastised before for not having discussed events that occurred only after my manuscript was finished, I would like to point readers to the blogosphere of the broadsheets (e.g. www.guardian.co.uk) and the BBC website (www.bbc.co.uk) for up to the minute commentary.
5. The term 'Asian' in British usage refers to immigrants from the Indian subcontinent. Other Asian migrants tend to be described in terms of their nationality or ethnic self-ascription.
6. Even after the new law was introduced, citizens of the Republic of Ireland living in the UK were regarded legally as citizens of another state but not as aliens/foreigners; among other reasons, this rather adventurous legal construct took account of the fact that, according to the laws of the Republic of Ireland, the inhabitants of Northern Ireland remained entitled to Irish citizenship.
7. One exception here is Travis (2002), who at least cited the relevant passage, albeit without giving much consideration to its implications.
8. Kivisto (2002: 163) cites the example of Bavaria, where ethnically segregated schools teach the children of immigrants in the language of their respective 'homeland' and so – disguised as multicultural practice – prepare them for returning 'home'. According to that same logic, the promotion of Welsh and

Gaelic may contribute – as a centrifugal force in multicultural disguise – to the dissolution of the UK.
9. See, for example, the letter by Simon Brooks to the editor of the *Guardian*, published on 17 September 2002.
10. The following paragraphs are based on a funding application that I wrote with Neringa Liubinienė in 2006 in connection with part of her doctoral fieldwork.

4 Third Journey – To the Market: *Trading Our Futures*

1. In 1989 the French National Geographic Institute calculated the precise location of the centre of Europe to be at 54° 51' North latitude and 25° 19' East longitude, near the village of Purnuškės, some 25 km north of Vilnius. Among the numerous centres of Europe that have been calculated in different ways, this is probably the most glamorous one since it was marked by a white column topped with a crown of golden stars officially unveiled on the day Lithuania joined the EU.
2. The term comprises a wide field relating to policymaking and the management of public affairs; it could also be translated as 'public administration' in a broad sense. An alternative German term often found in the literature on the history of ideas in European ethnology is *Allgemeine Statistik*, which refers in particular to early census and ordnance survey work.
3. Although a masculine noun, *Landsmann* derives from the archaic use of the term *Mann* (man) as synonymous with *Mensch* (human being). Similarly, *Mannschaft* refers to a group – usually a team – of people regardless of gender.
4. This section is based on a paper originally presented at the 33rd congress of the Deutsche Gesellschaft für Volkskunde at Jena in 2001. For their comments and constructive critique I am grateful especially to Klaus Roth, Péter Niedermüller, Alois Moosmüller and Irene Götz. Thanks are also due to participants in the seminar series 'Invitation to European Ethnology' at the University of the West of England, Bristol, especially to David Morley for his thought-provoking seminar, which helped to shape the first English draft version of this section, presented at the Vienna conference mentioned in the main text.
5. This is another case of subtle differences in translation. While 'hegemony' is commonly rendered in German as *Hegemonie* and vice versa, strictly speaking *Hegemonie* is a foreign loan word used to denote specifically what is captured by the more limited German term *Vorherrschaft* (predominance), that is, the *primus inter pares* of different, potentially conflicting kinds of *Herrschaft*. Gramsci's use of 'hegemony' seems to me to be closer to the German *Herrschaft* than to the English 'hegemony'; having read him, so far, only in English translations, I must reserve judgement on the matter.
6. There is apparently an interesting prehistory to the publication of Weisweiler's book. He hints that the manuscript had been ready since 1938 but could not be published until 1943. Although he does not say why, it may have had something to do with the way he contrasts the Celtic world view with that of imperial Rome – a contrast that might not have been politically

opportune to draw until after the 'axis' between Nazi Germany and Fascist Italy had collapsed (when it may even have become useful for propaganda purposes).

7. The term *gombeen* derives from the Irish (Gaelic) word for usury; the designation 'gombeen-man' is nowadays applied to men who (try to) direct local economic life in a semi-feudal manner.

8. Cited by Seamus Deane in *The Crane Bag* 8(1), 1984, 90.

9. The texts of all rulings are available on the OFT website (www.oft.gov.uk) via a keyword search for Tesco.

5 Fourth Journey – On the Grand Tour: *We Should Remember*

1. Earlier versions of this section were presented as a keynote address at an international conference to mark the launch of anthropology at the University of Klaipėda, Lithuania, in 2005 and as a seminar at the University of Stockholm in 2006. I am grateful for questions and comments, especially to Ulf Hannerz, Helena Wulff and Vytis Čiubrinskas.

2. This section is based on my inaugural lecture as Professor of Ethnology at the University of Ulster, delivered on 31 October 2007.

6 Fifth Journey – Towards Castalia: *To Re-Place Europe*

1. This section originated as a discussion paper for a symposium on the future of European ethnology, held at the University of Lund in 2007. I am grateful to Orvar Löfgren for inviting me to this symposium, and to him and the other participants, especially Regina Bendix, Jonas Frykman, Thomas Højrup and Reinhard Johler, for a stimulating discussion.

2. Kurasawa's book was published around the same time as the SIEF conference title was suggested, but it did not come to the attention of the programme committee until two years later. Coincidences such as this may lend support to the metaphysical notion that an idea whose time has come will always 'break through'.

3. This paragraph is my own adapted translation of the cover text of Pollack (2005).

4. Track 13: 'Penkios baltos gulbės' (Five white swans), *Ten ant Jūraċiu* (Land by the Sea), Druka, Klaipėda, no date.

5. One could say 'covered them' or 'tucked them in', but these literal translations would not capture the spirit of the original in the same way.

6. *Tutejsi* is the title of a film and community project promoting local traditional culture in the Polish-Belorussian borderlands by recording the old songs and stories, carried out in association with the Architecture Laboratory Żywej in Leipzig.

7. This section started life as 'Morphogenetic Fieldwork and the Anthropos: An Ethnological Meditation about a Dead Coyote on Rannoch Moor', a contribution to a symposium on Joseph Beuys and anthropology held at the University of Ulster on 26 April 2007. I would like to thank the participants

for their comments and criticisms, especially my Ph.D. student Victoria Walters, who organised the symposium as part of her doctoral research (Walters 2009).

8. European ethnologists have long seen things differently and were undertaking fieldwork 'at home' even at a time when colonial powers and their successors were generously funding anthropological research overseas. This is by no means to suggest that they were more advanced than their colleagues; it merely reflects their rather different morphogenetic disciplinary field. Closely related in the early modern period, the two fields drifted apart during the Age of Empires and have only recently begun to converge again. Other interpretations of their history are equally possible and entirely legitimate, but cannot be discussed here in depth.

9. The reader who detects in this discussion resonances of 'quantum entanglement' and a 'holographic paradigm' is entirely on the right track. The Ph.D. proposal mentioned earlier led to a thesis, in the course of which its author underwent a fieldwork-driven morphogenesis from economist through geographer to, after graduation, anthropologist and European ethnologist. The thesis contained the outline of a holography of social systems – a model of socio-economic development drawing extensively on theoretical physics, Gaia theory and ecosophy; see Kockel (1989b).

10. This is my translation of Riehl as quoted in Girtler (2004: 11): *Frei durch die Welt zu streifen, das Auge stets geöffnet für Natur und Volk ist eine lustige Arbeit, ein lustiges Spiel ist es nicht ... so rechne ich die Doppelarbeit des gleichzeitigen Wanderns und Forschens für besonders anstrengend, für anstrengender als das gründlichste Bücherstudium am Schreibtisch.*

11. This is my reading of Riehl as quoted in Girtler (2004: 11): *Nur der einsame, kunstgeübte Wanderer ... findet den raschen Blick und die nie erlahmende Spannkraft zum rastlosen Beobachten. ... Die besten Gedanken findet man immer dort, wo man unmittelbare Anschauung der Tatsachen gefunden hat, und die Gedanken wollen ... auch gleich frischweg erfasst und festgehalten sein.*

12. The term was coined by Hermann Bausinger, who established the Ludwig-Uhland-Institut at the University of Tübingen some five decades ago, to distinguish a sociologically reincarnated Volkskunde from other versions of cultural studies emerging at the time, which were based more on textual analysis and literary criticism. My hunch is that Beuys, with his interest in the material and substantive, would have felt much more at home with Bausinger's society-oriented approach than with some of the other forms of cultural studies that seem to focus increasingly on individual pathologies.

13. This has been explored, for example, in my earlier work on Beuys, in the context of a Tate Gallery Liverpool exhibition and conference titled 'The Revolution is Us' in 1993; see Kockel (1995).

14. Among the better-known visions that combine *anthropos*, *ethnos* and *topos* is that of Scottish polymath Patrick Geddes, who was a contemporary of that movement; for a discussion of his ideas, see, for example, Stephen (2004).

15. An eloquent summary of this perspective in the artist's own terms can be found in Beuys (1986).

16. While McIntosh does not explicitly refer to Buber, his title reflects the same inextricable link that Deffenbaugh (2006) identifies in Buber's thought between 'care of souls' and 'care of soils'.

17. 'Cultures' is yet another term that has become suspect because it apparently 'essentialises' cultural groups. As I am not aware of any viable alternative that adequately captures the actuality in question, I continue to use it.
18. In his 1974 action 'I Like America and America Likes Me', Beuys spent a week in a cage with a coyote. The action has been documented photographically, with an accompanying text, in a book originally published in 1976 (Tisdall 2008).
19. This curious naming of a place is also found in Ireland, where Kinlough, County Leitrim, is located at the foot of a lake that drains into Donegal Bay.

7   Envisioning the Not-Yet: *Before Coming Home*

1. Dictionary/thesaurus definitions of 'authentic' include valid, bona fide, reliable, dependable, realistic, accurate, faithful. There is no suggestion of exclusivity here; that suggestion is an addition the term has acquired in certain politically motivated discourses, including some putatively academic ones.

Bibliography

Adamson, I. (1982), *The Identity of Ulster: The Land, the Language and the People* (Bangor, ME: Pretani).

Adorno, T. (1973), *The Jargon of Authenticity* (Evanston, IL: Northwestern University Press).

Aitmatow, T. (1995), *Der weiße Dampfer* (Frankfurt/Main: Suhrkamp).

Alibhai-Brown, Y. (2000), *After Multiculturalism* (London: Foreign Policy Centre).

AlSayyad, N., ed. (2004), *The End of Tradition?* (London: Routledge).

Amery, C. (1979), *An den Feuern der Leyermark* (Munich: Heyne).

—— (1984), *Das Königsprojekt* (Munich: Heyne).

Anderson, B. (1983), *Imagined Communities* (London: Verso).

Anderson, M. and E. Bort, eds (1999), *The Irish Border: History, Culture, Politics* (Liverpool: Liverpool University Press).

Anttonen, P. (2005), *Tradition through Modernity: Postmodernism and the Nation-State in Folklore Scholarship* (Helsinki: Finnish Literature Society).

Appleton, J. (2002), 'Testing Britishness', in *Spiked*, 19 September <http://www.spiked-online.com/Articles/00000006DA58.htm> (accessed 25 August 2009).

Augé, M. and J. Colleyn (2006), *The World of the Anthropologist* (Oxford: Berg).

Åhlström, E., ed. (1999), *Cultural Itineraries in Rural Areas: Go Cultural with Pleiades!* (Llangollen: ECTARC).

Bade, K. (1994), *Homo Migrans: Wanderungen aus und nach Deutschland* (Essen: Klartext).

Balibar, É. (2004), *We, the People of Europe? Reflections on Transnational Citizenship* (Princeton, NJ: Princeton University Press).

Banton, M. (2001), 'National Integration in France and Britain', *Journal of Ethnic and Migration Studies* 27(1), 151–68.

Bartas, S., dir. (2004), *Children Lose Nothing*, Kinema 2003. DVD absolutmedien, Zentropa 2004.

Bassnett, S., ed. (1997), *Studying British Cultures: An Introduction* (London: Routledge).

Basso, K. (1996), *Wisdom Sits in Places: Landscape and Language among the Western Apache* (Albuquerque, NM: University of New Mexico Press).

Bauman, Z. (1999), *Culture as Praxis* (London: Sage).

—— (2004), *Europe: An Unfinished Adventure* (Cambridge: Polity).

Bausinger, H. (1990[1961]), *Folk Culture in a World of Technology* (Bloomington, IN: Indiana University Press).

—— (2000), *Typisch deutsch: Wie deutsch sind die Deutschen?* (Munich: Beck).

Bausinger, H., U. Jeggle, G. Korff and M. Scharfe (1993[1978]), *Grundzüge der Volkskunde* (Darmstadt: Wissenschaftliche Buchgesellschaft).

Beck, U. and E. Grande (2007), *Das kosmopolitische Europa: Gesellschaft und Politik in der Zweiten Moderne* (Frankfurt/Main: Suhrkamp).

Behnke, D. (2006), '"Europäische Vision": Creativity Contest for Young People – Closing Ceremony in Berlin', *Europe Direct Newsletter* 5, 3 June.

Bell, K., N. Jarman and T. Lefebvre (2004), *Migrant Workers in Northern Ireland* (Belfast: Institute for Conflict Research).

Bendix, R. (1997), *In Search of Authenticity: The Formation of Folklore Studies* (Madison, WI: University of Wisconsin Press).

—— (1999), 'Der Anthropologieladen: Plädoyer für eine Internationalisierung der Wissensproduktion', in Köstlin and Nikitsch (1999), 99–118.

Berlin, I. (1976), *Vico and Herder: Two Studies in the History of Ideas* (London: Hogarth).

Beuys, J. (1986), 'Talking About One's Own Country: Germany', *In Memoriam, Joseph Beuys: Obituaries, Essays, Speeches* (Bonn: Inter Nationes).

Bhabha, J. (1999), 'Belonging in Europe: Citizenship and Post-National Rights', *International Social Science Journal* 159, 11–23.

Biggs, I. (no date), *Volume 1: In Debatable Lands* (Bristol: Wild Conversations).

Bindemann, W., ed. (2001), *Strange Home Britain: Memories and Experiences of Germans in Britain* (Edinburgh: Alpha).

—— (2004), *... und manchmal umarmt vom Regen. Lebenswege und Lebensansichten von Deutschen in Schottland* (Berlin: Verbum/Selbstverlag).

Bloch, E. (1978[1959]), *Das Prinzip Hoffnung* (Frankfurt/Main: Suhrkamp).

Blunkett, D. (2001), *Secure Borders, Safe Haven: Integration and Diversity in Modern Britain* (London: Stationery Office).

—— (2002), 'Integration with Diversity: Globalisation and the Renewal of Democracy and Civil Society', in P. Griffith and M. Leonard, eds, *Reclaiming Britishness* (London: Foreign Policy Centre). <http://fpc.org.uk/articles/182> (accessed 25 August 2009).

Borgolte, M. (2005), 'Wie Europa seine Vielfalt fand: Über die mittelalterlichen Wurzeln für die Pluralität der Werte', in Hans Joas and Klaus Wiegandt, eds, *Die kulturellen Werte Europas* (Frankfurt/Main: Fischer), 117–63.

Bortoft, H. (1996), *The Wholeness of Nature: Goethe's Way of Science* (Edinburgh: Floris).

Boyle, D. and A. Simms (2009), *The New Economics: A Bigger Picture* (London: Earthscan).

Bragg, B. (2007), *The Progressive Patriot: A Search for Belonging* (London: Black Swan).

Breathnach, P. (1994), 'Gender and Employment in Irish Tourism', in U. Kockel, ed., *Culture, Heritage and Development: The Case of Ireland* (Liverpool: Liverpool University Press), 47–60.

Brednich, R., ed. (1988), *Grundriss der Volkskunde: Einführung in die Forschungsfelder der Europäischen Ethnologie* (Hamburg: Reimer).

Breidenbach, J. and I. Zukrigl (2000), *Tanz der Kulturen: Kulturelle Identität in einer globalisierten Welt* (Reinbek: Rowohlt).

Bremer, J. (1996), *Festschrift 150 Jahre Deutsche Kirche in Liverpool.* (Liverpool: Kirchenvorstand der Deutschen Kirche Liverpool).

Brett, D. (1996), *The Construction of Heritage* (Cork: Cork University Press).

Brewin, C. (1997), 'Society as a Kind of Community: Communitarian Voting with Equal Rights for Individuals in the European Union', in T. Modood and P. Werbner, eds, *The Politics of Multiculturalism in the New Europe: Racism, Identity and Community* (London: Zed), 223–39.

Briedis, L. (2008), *Vilnius: City of Strangers* (Vilnius: Baltos Lankos).

Brown, D. (1994), *The State and Ethnic Politics in Southeast Asia* (London: Routledge).

Buber, M. (1970), *I and Thou* (New York: Charles Scribner's Sons).

Burckhardt-Seebass, C. (1999), 'Die Verwissenschaftlichung des Selbsterlebten', in Köstlin and Nikitsch (1999), 119–26.

Burgess, A. (1997), *Divided Europe: The New Domination of the East* (London: Pluto).

Būgienė, L. (2005), 'Making Myths to Live by: Constructing Lithuanian Identity Against the Scandinavian Background', in Ronström and Palmenfelt, eds (2005), 126–50.

Byron, R. and U. Kockel, eds (2006), *Negotiating Culture: Moving, Mixing and Memory in Contemporary Europe* (Münster: LIT).

Caglar, A. (1997), 'Hyphenated Identities and the Limits of "Culture"', in Modood and Werbner (1997), 169–85.

Calvino, I. (1992), *If on a Winter's Night a Traveller* (London: Minerva).

Canetti, E. (1999), *The Tongue Set Free: Remembrance of a European Childhood* (London: Granta).

Capra, F. (1975), *The Tao of Physics: An Exploration of the Parallels between Modern Physics and Eastern Mysticism* (London: Fontana).

Carrier, J. and D. Miller, eds (1998), *Virtualism: A New Political Economy* (Oxford: Berg).

Carter, E., J. Donald and J. Squires, eds (1993), *Space and Place: Theories of Identity and Location* (London: Lawrence and Wishart).

Casey, E. (1998), *The Fate of Place* (Berkeley, CA: University of California Press).

Cederman, L. (2001), *Constructing Europe's Identity: The External Dimension* (London: Lynne Rienner).

Christiansen, P. (1996), 'Culture and Politics in a Historical Europe', *Ethnologia Europaea* 26(2): 137–46.

Christlich-Demokratische Union (2000), *Arbeitsgrundlage für die Zuwanderungs-Kommission der CDU Deutschlands*, 6 November. <http://www.cdu.de/doc/pdfc/1100_arbeitsgrundlage.pdf> (accessed 25 August 2009).

Clayton, P. (1996), *Enemies and Passing Friends: Settler Ideologies in Twentieth Century Ulster* (London: Pluto).

Clifford, J. and G. Marcus, eds (1986), *Writing Culture: The Poetics and Politics of Ethnography* (Berkeley, CA: University of California Press).

Coleman, S. and P. Collins, eds (2006), *Locating the Field: Space, Place and Context in Anthropology* (Oxford: Berg).

Coleman, W. (2004), *Economics and its Enemies: Two Centuries of Anti-Economics* (Basingstoke: Palgrave Macmillan).

Coles, A., ed. (2001), *Site-Specificity in Art: The Ethnographic Turn: 4 (de-, dis-, ex-)* (London: Black Dog).

Commission on the Future of Multi-Ethnic Britain (2000), *The Future of Multi-Ethnic Britain: The Parekh Report* (London: Profile).

Conroy, P. and A. Brennan (2003), *Migrant Workers and their Experiences* (Dublin: Equality Authority).

Cronin, A. (2002), 'Consumer Rights/Cultural Rights: A New Politics of European Belonging', *European Journal of Cultural Studies* 5(3), 307–23.

Čiubrinskas, V. (2000), 'Identity and the Revival of Tradition in Lithuania: An Insider's View', *Folk* 42, 19–40.

—— (2004), 'Transnational Identity and Heritage: Lithuania Imagined, Constructed and Contested, in Kockel and Nic Craith (2004), 42–66.

Dawe, G. and J. Foster, eds (1991), *The Poet's Place: Ulster Literature and Society. Essays in Honour of John Hewitt, 1907–1987* (Belfast: Institute of Irish Studies).

Day, R. (2000), *Multiculturalism and the History of Canadian Diversity* (Toronto, Ont.: University of Toronto Press).

DeCouflé, A. (1992), 'Historic Elements of the Politics of Nationality in France (1889–1989)', in D. Horowitz and G. Noiriel, eds, *Immigrants in Two Democracies: French and American Experiences* (New York: New York University Press), 357–67.

Deffenbaugh, D. (2006), *Learning the Language of the Fields: Tilling and Keeping as Christian Vocation* (Cambridge, MA: Cowley).

Delanty, G. (1995), *Inventing Europe: Idea, Identity, Reality* (London: Macmillan).

Demossier, M., ed. (2007), *The European Puzzle: The Political Structuring of Cultural Identities at a Time of Transition* (Oxford: Berghahn).

Denning, M. (2001), 'Globalization in Cultural Studies: Process and Epoch', *European Journal of Cultural Studies* 4(3), 351–64.

Devall, W. (1980), 'The Deep Ecology Movement', *Natural Resource Journal* 20, 299–322.

Dohmen, D. (1994), *Das deutsche Irlandbild. Imagologische Untersuchungen zur Darstellung Irlands und der Iren in der deutschssprachigen Literatur* (Amsterdam: Rodopi).

Donati, P. (1994), 'Towards a New European Citizenship – The Societal Idea', in Servizio Studi (1994), 197–245.

Dunlop, J. (2007), 'Language, Faith and Communication', in M. Nic Craith, ed., *Language, Power and Identity Politics* (Basingstoke: Palgrave Macmillan), 179–97.

Edwards, B. (1996), 'How the West was Wondered: County Clare and Directions in Irish Ethnography', *Folklore Forum* 27(2), 65–78.

Ehrlich R. and L. Luup, eds (1993), *Estonia: The New Tourist Destination* (Tallinn: Department of Tourism).

Ellis, P. (1988), *Hell or Connaught: Cromwellian Colonisation of Ireland, 1652–60* (Belfast: Blackstaff).

Elvert, J., ed. (1994), *Nordirland in Geschichte und Gegenwart / Northern Ireland – Past and Present* (Stuttgart: Steiner Wiesbaden).

Erdrich, L. (2003), *The Master Butcher's Singing Club* (London and New York: Harper Perennial).

Eriksen, A. (1997), 'Memory, History, and National Identity', *Ethnologia Europaea* 27(2), 129–37.

Europäische Visionen: 25 Filme, 25 Regisseure, Zentropa, 2004.

Europäischer Austausch gGmbH <www.european-exchange.org> (accessed 29 January 2008).

European Communities – Commission (1988), *A Fresh Boost for Culture in the European Community*, EC Bulletin Supplement 4/87 (Luxembourg: Office for Official Publications of the European Communities).

Eurovisionen – Vom kulturellen Netzwerk zur Politik <www.bpb.de/files/2QRNUI.pdf> (accessed 25 August 2009).

Evans, E. (1984), *Ulster: The Common Ground* (Mullingar: Lilliput).

—— (1992), *The Personality of Ireland: Habitat, Heritage and History* (Dublin: Lilliput).

Ferguson, R. (1988), *Chasing the Wild Goose: The Iona Community* (London: Fount).

Fitzpatrick, R. (1989), *God's Frontiersmen: The Scots-Irish Epic* (Chatswood, NSW: Peribo).

Fixico, D. (2003), *The American Indian Mind in a Linear World: American Indian Studies and Traditional Knowledge* (New York: Routledge).

Frevert, Ute (2003), *Eurovisionen: Ansichten guter Europäer im 19. und 20. Jahrhundert* (Frankfurt/Main: Fischer).

Friedlander, J. (1990), *Vilna on the Seine: Jewish Intellectuals in France Since 1968* (New Haven, CT: Yale University Press).

Friz, T. and E. Schmeckenbecher, eds (1979), *Es wollt ein Bauer früh aufstehn ... 222 Volkslieder* (Dortmund: Pläne).

Frykman, J. (1999), 'Belonging in Europe: Modern Identities in Minds and Places', *Ethnologia Europaea* 29(2), 13–24.

Frykman, J. and O. Löfgren (1987), *Culture Builders: A Historical Anthropology of Middle-Class Life* (New Brunswick, NJ: Rutgers University Press).

Frykman, J. and N. Gilje, eds (2003), *Being There: New Perspectives on Phenomenology and the Analysis of Culture* (Lund: Nordic Academic Press).

Fuhr, E. (2007), *Wo wir uns finden: Die Berliner Republik als Vaterland* (Berlin: BvT).

Fukuyama, F. (1992), *The End of History and the Last Man* (London: Penguin).

Gabanyi, A. (1992), 'Nationalismus in Rumänien: Vom Revolutionspatriotismus zur chauvinistischen Restauration', *Südosteuropa* 41(5), 275–92.

Gbadegesin, S. (2007), 'Reinventing the Yoruba Post-Obasanjo Presidency', *Nigerian Tribune* 19 September 2007.

Geiger, K., U. Jeggle and G. Korff, eds (1970), *Abschied vom Volksleben* (Tübingen: Tübinger Vereinigung für Volkskunde).

Geiss, I. (1993), *Europa – Vielfalt und Einheit: eine historische Erklärung*, Meyers Forum 12 (Mannheim: B.I.).

Gelfert, H. (2005), *Was ist deutsch? Wie die Deutschen wurden, was sie sind* (Munich: Beck).

Gell, A. (1998), *Art and Agency: An Anthropological Theory* (Oxford: Oxford University Press).

Gellner, E. (1983), *Nations and Nationalism* (Ithaca, NY: Cornell University Press).

Georgescu-Roegen, N. (1972), 'Economics and Entropy', *The Ecologist* 14, 194–200.

Gibbons, L. (1996), *Transformations in Irish Culture* (Cork: Cork University Press).

Giddens, A. (1999), 'Tradition', BBC Reith Lecture, 21 April. <http://news.bbc.co.uk/hi/english/static/events/reith_99/week3/week3.htm> (accessed 25 August 2009).

Gidoomal, R., D. Mahtani and D. Porter, eds (2001), *The British and How to Deal With Them: Doing Business with Britain's Ethnic Communities* (London: Middlesex University Press).

Girtler, R. (2004), *10 Gebote der Feldforschung* (Wien: LIT).

Glasman, M. (1996), *Unnecessary Suffering: Managing Market Utopia* (London: Verso).

Glassie, H. (1995[1982]), *Passing the Time in Ballymenone: Culture and History of an Ulster Community* (Bloomington, IN: Indiana University Press).

—— (2006), *The Stars of Ballymenone* (Bloomington, IN: Indiana University Press).

Görner, L., ed. (no date), *Heinrich Heine (1797–1856): Ein Lesebuch für Demokraten, und solche, die es werden wollen* (self-published by the editor).

Gorjanicyn, K. (2000), 'Citizenship and Culture in Contemporary France: Extreme Right Interventions', in A. Vandenberg, ed., *Citizenship and Democracy in a Global Era* (Basingstoke: Palgrave Macmillan), 138–55.

Graeber, D. (2001), *Toward an Anthropological Theory of Value: The False Coin of Our Own Dreams* (New York: Palgrave Macmillan).

Graham, B. (1997), *In Search of Ireland: A Cultural Geography* (London: Routledge), 192–212.

Greverus, I. (1979), *Auf der Suche nach Heimat* (München: Beck).
—— (1990), 'Anthropological Horizons, the Humanities and Human Practice', *Anthropological Journal on European Cultures* 1(1), 13–33.
—— (2005), *Ästhetische Orte und Zeichen: Wege zu einer ästhetischen Anthropologie* (Münster: LIT).
Grimes, S. (1988), 'The Sydney Irish: A Hidden Ethnic Group', *Irish Geography* 21(2): 69–77.
Gudeman, S. (2008), *Economy's Tension: The Dialectics of Community and Market* (New York: Berghahn).
Guildhall Press (2008), *Murals of Derry* (Derry: Guildhall).
Hale, A. (2002), 'Creating a Cornish Brand: Discourses of "Traditionality" in Cornish Regeneration Strategies', in Kockel, ed. (2002), 164–74.
Hall, S. (1995), 'New Cultures for Old', in D. Massey and P. Jess, eds, *A Place in the World: Places, Cultures and Globalization* (Oxford: Oxford University Press), 175–214.
Haller, M. (1994), 'Ethnic and National Identities and Their Relations in the New European Scenario', in Servizio Studi (1994), 69–89.
Handwerker, P. (1997), 'Universal Human Rights and the Problem of Unbounded Cultural Meanings', *American Anthropologist* 99(4): 799–809.
Hann, C. (2006), *'Not the Horse We Wanted!' Postsocialism, Neoliberalism, and Eurasia* (Münster: LIT).
Hannerz, U. (1996), 'Flows, Boundaries and Hybrids: Keywords in Transnational Anthropology', plenary lecture at the twentieth biennial meeting of the Associacao Brasileira de Antropologia at Salvador de Bahia, 14–17 April.
—— (1997), 'Borders', *International Social Science Journal* 154, 537–48.
—— (2006), 'Studying Down, Up, Sideways, Through, Backwards, Forwards, Away and at Home: Reflections on the Field Worries of an Expansive Discipline', in Coleman and Collins (2006), 23–41.
Hardt, M. and A. Negri (2000), *Empire* (Cambridge, MA: Harvard University Press).
—— (2004), *Multitude: War and Democracy in the Age of Empire* (New York: Penguin).
Harris, R. (1972), *Prejudice and Tolerance in Ulster: A Study of 'Neighbours' and 'Strangers' in a Border Community* (Manchester: Manchester University Press).
Harte, L., Y. Whelan and P. Crotty, eds (2005), *Ireland: Space, Text, Time* (Dublin: Liffey).
Hartmann, A. (1988), 'Die Anfänge der Volkskunde', in Brednich (1988), 9–30.
Hartung, W. (1991), '"Das Vaterland als Ort der Heimat". Grundmuster konservativer Identitätsstiftung und Kulturpolitik in Deutschland', in Klueting (1991), 112–56.
Harvey, D. (1989), *The Condition of Postmodernity* (Oxford: Blackwell).
Hecht, M. (2000), *Das Verschwinden der Heimat: Zur Gefühlslage der Nation* (Leipzig: Reclam).
Hechter, M. (1975) *Internal Colonialism: The Celtic Fringe in British National Development, 1536–1966* (London: Routledge & Kegan Paul).
Heelas, P. (1996), 'Introduction: Detraditionalization and its Rivals', in Heelas, Lash and Morris (1996), 1–20.
Heelas, P., S. Lash and P. Morris, eds (1996), *Detraditionalization: Critical Reflections on Authority and Identity* (Oxford: Blackwell).
Hengartner, T. (2001), '"Kulturwissenschaftler sind wir alle". Selbstbehauptung und Selbstverständnis eines (kleinen) Faches in einer leistungs- und marktorientierten Hochschullandschaft', in König and Korff (2001), 39–50.

Heslinga, M. (1962), *The Irish Border as a Cultural Divide: A Study of Regionalism in the British Isles* (Assen: van Gorcum).

Hirsch, F. (1977), *The Social Limits to Growth* (London: Routledge and Kegan Paul).

Hobsbawm, E. (1992[1983]), 'Introduction: Inventing Traditions', in E. Hobsbawm and T. Ranger, eds, *The Invention of Tradition* (Cambridge: Cambridge University Press), 1–14.

Højrup, T. (2003), *State, Culture and Life-Modes: The Foundations of Life-Mode Analysis* (Aldershot: Ashgate).

Holy, L. and M. Stuchlik (1983), *Actions, Norms and Representations: Foundations of Anthropological Enquiry* (Cambridge: Cambridge University Press).

Hylland, O. (2001), 'On the Study of "Folk" as a Category: The Popular in Popular Enlightenment', in U. Wolf-Knuts and A. Kaivola-Bregenhøj, eds, *Pathways: Approaches to the Study and Teaching of Folklore* (Turku: Nordic Network of Folklore), 17–25.

Ifversen, J. (2002), 'Europe and European Culture – A Conceptual Analysis', *European Societies* 4(1), 1–26.

Ivakhiv, A. (2006), 'Stoking the Heart of (a Certain) Europe: Crafting Hybrid Identities in the Ukraine-EU Borderlands', *Spaces of Identity* 6(1), 11–44.

Jackson, A., ed. (1987), *Anthropology at Home* (London: Tavistock).

Jacobson, D. (1996), *Rights across Borders: Immigration and the Decline of Citizenship* (Baltimore, MD: Johns Hopkins University Press).

James, H. (1991), *Deutsche Identität. 1770–1990* (Frankfurt/Main: Campus).

Jarman, N. (2005), *Changing Patterns and Future Planning: Migration and Northern Ireland*, ICR Working Paper 1, Belfast, December.

—— (2006), 'Diversity, Economy and Policy: New Patterns of Migration to Northern Ireland', *Shared Space* 2, 45–61.

Jeggle, U., G. Korff, M. Scharfe and B. Warneken, eds (1986), *Volkskultur in der Moderne: Probleme und Perspektiven empirischer Kulturforschung* (Reinbek: Rowohlt).

Jochum, K. (2006), 'Yeats in Germany, Austria and Switzerland', in K. Jochum, ed., *The Reception of W. B. Yeats in Europe* (London: Continuum), 50–75.

Johler, R. (1999a) '"Europa in Zahlen": Statistik – Vergleich – Volkskunde – EU' *Zeitschrift für Volkskunde* 95, 246–63.

—— (1999b), 'Telling a National Story with Europe: Europe and the European Ethnology', *Ethnologia Europaea* 29(2), 67–74.

(2002), 'The EU as Manufacturer of Tradition and Cultural Heritage', in Kockel, ed. (2002), 221–30.

Joseph, A. (2001), *The People's Gallery* (Derry: Bogside Artists).

Jung, C. (1991), *The Archetypes and the Collective Unconscious*, 2nd edn (London: Routledge).

Kamberger, K. (1981), *Mit dem Hintern am Boden und dem Kopf in den Wolken: Entdeckungsfahrten Richtung Heimat* (Frankfurt/Main: Eichborn).

Kan, S. and Pauline Turner Strong, eds (2006), *New Perspectives on Native North America: Cultures, Histories, and Representations* (Lincoln, NE: University of Nebraska Press).

Kapuściński, R. (2008), *Der Andere* (Frankfurt/Main: Suhrkamp).

Kaser, K., D. Gramshammer-Hohl and R. Pichler, eds (2004), *Europa und die Grenzen im Kopf* (Klagenfurt: Wieser).

Katwala, S. (2001), 'The Truth of Multicultural Britain', *Observer on Sunday*, 25 November.

Katzenstein, P., ed. (1997), *Mitteleuropa: Between Europe and Germany* (Oxford and New York: Berghahn).

Kennard, A. (2000), 'The Role of Central and Eastern Europe in the EU's Regional Planning Agenda for the New Millennium', *Journal of European Area Studies* 8(2), 203–19.

Kettenacker, L. (1996), 'The Germans after 1945', in Panayi (1996), 187–208.

Khan, S. (2001), 'Devolved and Diverse?' *The Observer on Sunday*, 25 November.

Kiberd, D. (2005), *The Irish Writer and the World* (Cambridge: Cambridge University Press).

Kidwell, C. and A. Velie (2005), *Native American Studies* (Edinburgh: Edinburgh University Press).

Kilday, A., ed. (1998), *Culture and Economic Development in the Regions of Europe* (Llangollen: ECTARC).

Kimmerle, H. (2002), *Interkulturelle Philosophie: Zur Einführung* (Hamburg: Junius).

Kivisto, P. (2002), *Multiculturalism in a Global Society* (Oxford: Blackwell).

Klueting, E., ed. (1991), *Antimodernismus und Reform: Zur Geschichte der deutschen Heimatbewegung* (Darmstadt: Wissenschaftliche Buchgesellschaft).

Klüter, H. (1986), *Raum als Element sozialer Kommunikation*, Giessener Geographische Schriften 60, University of Giessen.

Klusmeyer, D. (2001), 'A "Guiding Culture" for Immigrants? Integration and Diversity in Germany', *Journal of Ethnic and Migration Studies* 27(3), 519–32.

Kockel, U. (1989a), 'Immigrants – Entrepreneurs of the Future?' *Common Ground* 70, 6–8.

—— (1989b), *Political Economy, Everyday Culture, and Change: A Study of Informal Economy and Regional Development in the West of Ireland* (Ph.D. thesis, University of Liverpool).

—— (1991), 'Countercultural Migrants in the West of Ireland', in R. King, ed., *Contemporary Irish Migration* (Dublin: Geographical Society of Ireland), 70–82.

—— (1992), 'Provisory Economy and Regional Development: Towards a Conceptual Integration of Informal Activities', in M. Tykkyläinen, ed., *Development Issues and Strategies in the New Europe: Local, Regional and Interregional Perspectives* (Aldershot: Avebury), 101–19.

—— (1994), 'Mythos und Identität. Der Konflikt im Spiegel der Volkskultur', in Elvert (1994), 495–517.

—— (1995), 'The Celtic Quest: Beuys as Hero and Hedge School Master', in D. Thistlewood, ed., *Joseph Beuys: Diverging Critiques* (Liverpool: Tate Gallery and Liverpool University Press), 129–47.

—— (1999a), *Borderline Cases: The Ethnic Frontiers of European Integration* (Liverpool: Liverpool University Press).

—— (1999b), 'Nationality, Identity, Citizenship: Reflections on Europe at Drumcree Parish Church', *Ethnologia Europaea* 29(2), 97–108.

—— (2001a), 'Geopolitik und kulturelles Erinnern in Nordirland. Versuch einer virtuellen Geschichte', *kuckuck. Notizen zur Alltagskultur* 17(1), 42–7.

—— (2001b), 'Protestantische Felder in katholischer Wildnis. Zur Politisierung der Kulturlandschaft in Ulster', in R. Brednich, A. Schneider and U. Werner, eds, *Natur – Kultur: Volkskundliche Perspektiven auf Mensch und Umwelt* (Münster: Waxmann), 125–38.

—— (2002a), *Regional Culture and Economic Development: Explorations in European Ethnology* (Aldershot: Ashgate).

——, ed. (2002b), *Culture and Economy: Contemporary Perspectives* (Aldershot: Ashgate).

—— (2003a), 'EuroVisions: Journeys to the Heart of a Lost Continent', *Journal of Contemporary European Studies* 11(1), 53–66.

—— (2003b), 'Heimat als Widerständigkeit: Beobachtungen in einem Europa freischwebender Regionen', in S. Götsch and C. Köhle-Hezinger, eds, *Komplexe Welt: Kulturelle Ordnungssysteme als Orientierung* (Münster: Waxmann), 167–76.

—— (2004), 'Von der Schwierigkeit, "Britisch" zu sein. Monokulturelle Politik auf dem Weg zur polykulturellen Gesellschaft', in C. Köck, A. Moosmüller and K. Roth, eds, *Zuwanderung und Integration: Kulturwissenschaftliche Zugänge und soziale Praxis* (Münster: Waxmann), 65–81.

—— (2005a), '"Authentisch ist, was funktioniert!" Tradition und Identität in drei irischen Städten', in S. Göttsch, W. Kaschuba and K. Vanja, eds, *Ort – Arbeit – Körper: Ethnografie europäischer Modernen* (Münster: Waxmann), 127–34.

—— (2005b), 'Frontiers', in G. Welz and R. Lenz, eds, *Von Alltagswelt bis Wandmalerei: Eine kleine Enzyklopädie. Ina-Maria Greverus zum Fünfundsiebzigsten* (Münster: LIT), 62–3.

—— (2007a), 'Heritage versus Tradition: Cultural Resources for a New Europe?' in Demossier (2007), 85–101.

—— (2007b), 'K(l)eine Deutschlande. Heimat und Fremde deutscher Einwanderer auf den Britischen Inseln', in E. Tschernokoshewa and V. Granssow, eds, *Beziehungsgeschichten: Minderheiten – Mehrheiten in europäischer Perspektive* (Münster: Waxmann), 188–202.

—— (2007c), 'Reflexive Traditions and Heritage Production', in Nic Craith and Kockel (2007), 19–33.

—— (2008a), 'Editorial', *Anthropological Journal of European Cultures* 17(1), 1–4.

—— (2008b), 'Liberating the Ethnological Imagination', *Ethnologia Europaea* 38(1), 8–12.

—— (2008c), 'Putting the Folk in Their Place: Tradition, Ecology, and the Public Role of Ethnology', *Anthropological Journal of European Cultures* 17(1), 5–23.

—— (2008d), 'Turning the World Upside Down: Towards a European Ethnology in and of England', in Nic Craith, Kockel and Johler (2008), 149–63.

—— (2009), 'Wozu eine Europäische Ethnologie – und welche?' *Österreichische Zeitschrift für Volkskunde* LXIII/112(3), 39–56.

Kockel, U. and M. Nic Craith, eds (2004), *Communicating Cultures* (Münster: LIT).

König, G. and G. Korff, eds (2001), *Volkskunde '00: Hochschulreform und Fachidentität* (Tübingen: TVV).

Köstlin, K. (1996), 'Perspectives of European Ethnology', *Ethnologia Europaea* 26(2), 169–80.

—— (1999), 'Ethnographisches Wissen als Kulturtechnik', in Köstlin and Nikitsch (1999), 9–30.

Köstlin, K. and H. Nikitsch, eds (1999), *Ethnographisches Wissen: Zu einer Kulturtechnik der Moderne* (Wien: Institut für Volkskunde).

Koopmans, R. and P. Statham (1998), 'Challenging the Liberal Nation-State? Postnationalism, Multiculturalism, and the Collective Claims making of Migrants and Ethnic Minorities in Britain and Germany', *American Journal of Sociology* 105, 652–96.

Kosuth, J. (1991), *Art after Philosophy and after: Collected Writings 1966–1990* (Cambridge, MA: MIT Press).

Kramer, D. (1997), *Von der Notwendigkeit der Kulturwissenschaften: Aufsätze zu Volkskunde und Kulturtheorie* (Marburg: Jonas).

Krasnodebski, Z. (1994), 'Universalism and Pluralism in European Culture and Cultural Problems within Central Eastern Europe', in Servizio Studi (1994), 37–54.

Kristmundsdottir, S. (2006), 'Far from the Trobriands? Biography as Field', in Coleman and Collins (2006), 163–77.

Krockow, C. (1992), *Heimat: Erfahrungen mit einem deutschen Thema* (München: Deutscher Taschenbuchverlag).

Kumar, K. (2001), '"Englishness" and English National Identity', in Morley and Robins (2001), 41–55.

Kundera, M. (1990), *Croí na hEorpa* (Baile Átha Cliath: Coiscéim).

Kurasawa, F. (2004), *The Ethnological Imagination: A Cross-Cultural Critique of Modernity* (Minneapolis, MN: University of Minnesota Press).

Kymlicka, W. (1995), *Multicultural Citizenship: A Liberal Theory of Minority Rights* (New York: Clarendon).

Law, A. (2005), 'The Ghost of Patrick Geddes: Civics as Applied Sociology', *Journal of Generalism & Civics* VI, 4–19. <http://patrickgeddes.co.uk/generalism_civics_six.pdf> (accessed 25 August 2009).

Lenclud, G. (2003), 'Tradition is No Longer What it was ... On the Notions of Tradition and Traditional Societies in Ethnology', in L. Varadarajan and D. Chevalier, eds, *Tradition and Transmission: Current Trends in French Ethnology. The Relevance for India* (New Delhi: Aryan), 72–93.

Lepenies, W. (2006), *Kultur und Politik: Deutsche Geschichten* (München: Hanser).

Lerm Hayes, C. (2006), 'Unity in Diversity through Art? Joseph Beuys' Models of Cultural Dialogue', paper presented at the first EURODIV Conference, 'Understanding Diversity', Milan/Italy, 26–27 January. <http://www.feem.it/NR/rdonlyres/C47F6623-18D1-4BA1-8706-C5E80A694A2E/1968/6008.pdf> (accessed 8 April 2009).

Liubinienė, N. (2008), 'Lithuanians in Northern Ireland: New Home, New Homeland?' *Irish Journal of Anthropology* 11(1), 9–13.

—— (2009), *Migrantai iš Lietuvos Šiaurės Airijoje: "Savos Erdvės" Konstravimas* (Ph.D. thesis, Vytautas Magnus University Kaunas).

Löfgren, O. (1987), 'Rational and Sensitive: Changing Attitudes to Time, Nature, and the Home', in Frykman and Löfgren (1987), 13–153.

—— (1996), 'Linking the Local, the National and the Global: Past and Present Trends in European Ethnology', *Ethnologia Europaea* 26(2), 157–68.

—— (2001), 'Life after Postmodernity: Volkskunde in the New Economy', in König and Korff (2001), 151–62.

Loftus, B. (1994), *Mirrors – Orange & Green* (Dundrum: Picture).

Lornell, C. and T. Mealor (1983), 'Traditions and Research Opportunities in Folk Geography', *Professional Geographer* 35(1), 51–6.

Maase, K. (1998), 'Nahwelten zwischen "Heimat" und "Kulisse": Anmerkungen zur volkskundlich-kulturwissenschaftlichen Regionalitätsforschung', *Zeitschrift für Volkskunde* 94, 53–70.

Maddox, J. (1981), Editorial: 'A Book for Burning?' *Nature* 293, 24 September, 245–6.

Mall, R. (2000), *Mensch und Geschichte: Wider die Anthropozentrik* (Darmstadt: Wissenschaftliche Buchgesellschaft).

Mallory J. and T. McNeill (1991), *The Archaeology of Ulster: From Colonization to Plantation* (Belfast: Institute of Irish Studies).

Malmborg, M. and B. Stråth, eds (2002), *The Meaning of Europe: Variety and Contention Within and Among Nations* (Oxford: Berg).

Massey, D. (1994), *Space, Place and Gender* (Cambridge: Polity).

McCrone, D. (2001), 'Scotland and the Union: Changing Identities in the British State', in Morley and Robins (2001), 97–108.

McCurdy, M. and A. Murphy, eds (1997), *Evangelisch-Lutherische Kirche in Irland 1697–1997* (Belfast: Evangelisch-Lutherische Kirche in Irland).

McDermott, P. (2008), 'Towards Linguistic Diversity? Community Languages in Northern Ireland', *Shared Space* 5, 5–20.

McFarlane, B. (1994), 'Is European Integration qualified by a New Balkanisation? Some Economic Aspects', *History of European Ideas* 19(1–3), 519–25.

McIntosh, A. (1998), 'Deep Ecology and the Last Wolf', *The Aisling Magazine* 23. <http://www.aislingmagazine.com/aislingmagazine/articles/TAM23/Deep.html> (accessed 25 August 2009).

—— (2002), *Soil and Soul: People versus Corporate Power* (London: Aurum).

McManus, T. (2007), *The Radical Field: Kenneth White and Geopoetics* (Dingwall: Sandstone).

Meek, D., ed. (1995), *Tuath is Tighearna: Tenants and Landlords: An Anthology of Gaelic Poetry of Social and Political Protest from the Clearances to the Land Agitation (1800–1890)* (Edinburgh: Gaelic Texts Society and Scottish Academic Press).

Melotti, U. (1997), 'International Migration in Europe: Social Projects and Political Cultures', in T. Modood and P. Werbner, eds (1997), 73–92.

Merleau-Ponty, M. (1960), 'De Mauss à Lévi-Strauss', *Signes*, 143–57.

Milton, K. (1996), *Environmentalism and Cultural Theory: Exploring the Role of Anthropology in Environmental Discourse* (London: Routledge).

Miłosz, C. (2002), *Native Realm: A Search for Self-Definition* (New York: Farrar, Straus and Giroux).

Modood, T. (1992), *Not Easy Being British: Colour, Culture and Citizenship* (Stoke-on-Trent: Trentham).

—— (2001), 'British Asian Identities: Something Old, Something Borrowed, Something New', in Morley and Robins (2001), 67–78.

—— (2002), 'Sources of Muslim Assertiveness in Britain: A Response to Alison Shaw', *Anthropology Today* 18(2), 24–5.

Moosmüller, A. (2000), 'Perspektiven des Fachs interkulturelle Kommunikation aus kulturwissenschaftlicher Sicht', *Zeitschrift für Volkskunde* 96, 169–85.

Morley, D. (2000), *Home Territories: Media, Mobility and Identity* (London and New York: Routledge).

Morley, D. and K. Robins (1995), *Spaces of Identity: Global Media, Electronic Landscapes and Cultural Boundaries* (London: Routledge).

——, eds (2001), *British Cultural Studies: Geography, Nationality and Identity* (Oxford: Oxford University Press).

Morphy, H. and M. Perkins, eds (2005), *The Anthropology of Art: A Reader* (Oxford: Blackwell).

Morris, L. (2002), 'Britain's Asylum and Immigration Regime: The Shifting Contours of Rights', *Journal of Ethnic and Migration Studies* 28, 409–25.

Morrow, D. (1994a), 'Games between Frontiers: Northern Ireland as Ethnic Frontier', in Elvert (1994), 334–53.

—— (1994b), 'Faith and Fervour: Religion and Nationality in Ulster', in Elvert (1994), 422–41.

Moxnes, H. (2003), *Putting Jesus in His Place: A Radical Vision of Household and Kingdom* (Louisville, KY: Westminster John Knox Press).

Nabokov, P. (2007), *Where the Lightning Strikes: The Lives of American Indian Sacred Places* (New York: Penguin).

Naess, A. (1973), 'The Shallow and the Deep, Long Range Ecology Movement: A Summary', *Inquiry* 16, 95–100.

Nagel, C. (2001), 'Hidden Minorities and the Politics of "Race": The Case of British Arab Activists in London', *Journal of Ethnic and Migration Studies* 27(3), 381–400.

Nelde, P., M. Strubell and G. Williams (1996), *Euromosaic: The Production and Reproduction of the Minority Language Groups of the EU* (Luxembourg: Office for Official Publications of the European Communities).

Nic Craith, M. (2000), 'Contested Identities and the Quest for Legitimacy', *Journal of Multilingual and Multicultural Development* 21(5), 399–413.

—— (2001), 'Politicised Linguistic Consciousness: The Case of Ulster-Scots', *Nations and Nationalism* 7(1), 21–37.

—— (2002), *Plural Identities – Singular Narratives: The Case of Northern Ireland* (Oxford: Berghahn).

—— (2003), *Culture and Identity Politics in Northern Ireland* (Basingstoke: Palgrave Macmillan).

—— (2004a), 'Conceptions of Equality: The Case of Northern Ireland', in A. Finlay, ed., *Nationalism and Multiculturalism: Irish Identity, Citizenship and the Peace Process* (Münster: LIT), 111–30.

—— (2004b), 'Culture and Citizenship in Europe: Questions for Anthropologists', *Social Anthropology* 12, 289–300.

—— (2007), *Europe and the Politics of Language* (Basingstoke: Palgrave Macmillan).

—— (2008a), 'From National to Transnational: A Discipline *en route* to Europe', in Nic Craith, Kockel and Johler (2008), 1–17.

—— (2008b), 'Intangible Cultural Heritages: The Challenge for Europe', *Anthropological Journal of European Cultures* 18(1), 54–73.

Nic Craith, M., ed. (1996), *Watching One's Tongue: Issues in Language Planning* (Liverpool: Liverpool University Press).

Nic Craith, M. and U. Kockel, eds (2007), *Cultural Heritages as Reflexive Traditions* (Basingstoke: Palgrave Macmillan).

Nic Craith, M., U. Kockel and R. Johler, eds (2008), *Everyday Culture in Europe: Approaches and Methodologies* (Aldershot: Ashgate).

Niedermüller, P. (1994), 'Politics, Culture and Social Symbolism: Some Remarks on the Anthropology of Eastern European Nationalism', *Ethnologia Europaea* 24(1), 21–33.

—— (1997), 'Postsozialismus, Kultur und Alltag im "wilden Osten": Zur kulturellen Repräsentation Osteuropas', Inaugural Lecture at the Humboldt-Universität zu Berlin, January 1997.

—— (1999), 'Ethnographie Osteuropas. Wissen, Repräsentation, Imagination: Thesen und Überlegungen', in Köstlin and Nikitsch (1999), 42–67.

Nogués, A. (2002), 'Culture, Transactions, and Profitable Meanings: Tourism in Andalusia', in Kockel, ed. (2002), 147–63.

O'Connor, P. (1989), *People Make Places: The Story of the Irish Palatines* (Newcastle West: Oireacht na Mumhan).

O'Donohue, J. (2003), *Divine Beauty: The Invisible Embrace* (London: Bantam).

Office for National Statistics (2006), *International Migration: Migrants Entering or Leaving the United Kingdom and England and Wales, 2004* (London: Office for National Statistics).

Olbracht, I. (1999), *The Sorrowful Eyes of Hannah Karajich* (London: Harvill 1999).

Oommen, T. (1994), 'Intercultural Communication in Multinational Settings – The Case of the European Community', in Servizio Studi (1994), 171–96.

O'Sullivan, N. (2007), 'Visions of European Unity since 1945', Elie Kedourie Memorial Lecture, British Academy, London, 17 May 2007.

Otto, K. (1977), *Vom Ostermarsch zur APO: Geschichte der außerparlamentarischen Opposition in der Bundesrepublik 1960–70* (Frankfurt/Main: Campus).

Panayi, P. (1996), 'Germans in Britain's History', in Panayi, ed., *Germans in Britain since 1500* (London: Hambledon), 1–16.

Panteļējevs, A. (1991), 'Die Scharfe Waffe der Dialektik. Der sowjetische Internationalismus in Theorie und Praxis', in Urdze (1991), 65–75.

Partridge, S. (1999), *The British Union State: Imperial Hangover or Flexible Citizens' Home?* Catalyst pamphlet 4, London.

Petersson, B. and E. Clark, eds (2003), *Identity Dynamics and the Construction of Boundaries* (Lund: Nordic Academic).

Pietersee, J. (1999), 'Europe, Travelling Light: Europeanization and Globalization', *The European Legacy* 4(3), 3–17.

Pines, J. (2001), 'Rituals and Representations of Black "Britishness"', in Morley and Robins (2001), 57–66.

Plessner, H. (1982[1959]), *Die verspätete Nation: Über die politische Verführbarkeit bürgerlichen Geistes* (Frankfurt/Main: Suhrkamp).

Pollack, M., ed. (2005), *Sarmatische Landschaften: Nachrichten aus Litauen, Belarus, der Ukraine, Polen und Deutschland* (Frankfurt/Main: Fischer).

Pratt, S. (2001), 'The Given Land: Black Hawk's Conception of Place', *Philosophy & Geography* 4(1), 109–25.

Price, A., C. Ó. Torna and A. Wynne Jones (1997), *The Diversity Dividend: Language, Culture and Economy in an Integrated Europe* (Brussels: European Bureau for Lesser Used Languages).

Purchla, J., ed. (2001), *From the World of Borders to the World of Horizons* (Krakow: International Cultural Centre).

Rainbird, S. (2005), *Joseph Beuys and the Celtic World: Scotland, Ireland and England 1970–85* (London: Tate).

Rasche, H. (1995), '"A Dacent and Quiet People": Palatine Settlements in County Limerick', in U. Kockel, ed., *Landscape, Heritage and Identity: Case Studies in Irish Ethnography* (Liverpool: Liverpool University Press), 155–77.

Rathgeb, E. (2005), *Die engagierte Nation: Deutsche Debatten 1945–2005* (München: Hanser).

Ray, C. (1998), 'Culture, Intellectual Property and Territorial Rural Development', *Sociologia Ruralis* 38(1), 3–20.

Reisenleitner, M. (2001), 'Tradition, Cultural Boundaries and the Constructions of Spaces of Identity', *Spaces of Identity* 1(1), 7–13.

Reulecke, J. (1991), 'Wo liegt Falado? Überlegungen zum Verhältnis von Jugendbewegung und Heimatbewegung vor dem Ersten Weltkrieg', in Klueting (1991), 1–19.

Riehl, W. (1903[1861]), *Land und Leute 2: Wanderbuch*, Stuttgart, quoted in Roland Girtler (2004), 11.

Ritchie, J. (1996), 'German Refugees from Nazism', in Panayi (1996), 147–70.

Rizzardo, R. (1987), *Cultural Policy and Regional Identity in Finland: North Karelia between Tradition and Modernity* (Strasbourg: Council for Cultural Co-operation).

Rogers, C. (1997), 'Explorations in Terra Incognita', *American Anthropologist* 99, 717–9.

Roll, E. (1978), *A History of Economic Thought*, 4th edn (London: Faber & Faber).

Ronström, O. (2005), 'Memories, Tradition, Heritage', in Ronström and Palmenfelt, eds (2005), 88–106.

Ronström, O. and U. Palmenfelt, eds (2005), *Memories and Visions* (Tartu: Tartu University Press).

Rosecrance, R. (2005), 'Europa gehört die Zukunft: Amerika ist altmodisch, Europa modern. Sie brauchen einander', *Internationale Politik* 60(7), 52–6.

Rosenkranz, A. (1965), *Geschichte der Deutschen Evangelischen Kirche in Liverpool* (Düsseldorf: Presseverband der Evangelischen Kirche im Rheinland).

Roth, K. (1996), 'European Ethnology and Intercultural Communication', *Ethnologia Europaea* 26(1), 3–16.

Schama, S. (1996), *Landscape and Memory* (London: Fontana).

Scharfe, M. (2001), '"Nicht das Knien hilft beim Beten, aber man kniet." Überlegungen zur volkskundlichen Fachidentität', in König and Korff (2001), 59–69.

Schiffauer, W. (1996), 'The Fear of Difference: New Trends in Cultural Anthropology', *Anthropological Journal on European Cultures* 5(1), 49–62.

—— (1997), *Fremde in der Stadt. Zehn Essays über Kultur und Differenz* (Frankfurt/Main: Suhrkamp).

Schlink, B. (2000), *Heimat als Utopie* (Frankfurt/Main: Suhrkamp).

Schlögel, K. (2002), *Die Mitte liegt ostwärts: Europa im Übergang* (Frankfurt/Main: Fischer).

Schmidt, A., ed. (1994), *Die Kurische Nehrung: Ostpreußische Dichter erzählen* (Husum: Husum).

Schmidt, B. (1994), *Am Jenseits zur Heimat. Gegen die Utopiefeindlichkeit im Dekonstruktiven* (Darmstadt: Wissenschaftliche Buchgesellschaft).

Schneider, A. and C. Wright, eds (2005), *Contemporary Art and Anthropology* (Oxford: Berg).

Schulze, H. (1990), *Die Wiederkehr Europas* (Berlin: Siedler).

Schwegler, T. (2009), 'The Global Crisis of Economic Meaning', *Anthropology News* 50(7), 9, 12.

Segert, D. (2002), *Die Grenzen Osteuropas: 1918, 1945, 1989 – Drei Versuche im Westen anzukommen* (Frankfurt/Main: Campus).

Sennett, R. (1996), 'The Foreigner', in Heelas, Lash and Morris (1996), 173–99.

Servizio Studi of the Autonomous Region of Trentino-South Tyrol, eds (1994): *Regional Diversity in Europe: The Role of Different Cultures in the Construction of the European Union. Conference Proceedings, Trento, October 1st, 1993* (Bolzano: Servizio Studi).

Shaw, A. (2002), 'Why Might Young British Muslims Support the Taliban?' *Anthropology Today* 18(1), 5–8.

Sheeran, P. (1988a), 'Genius Fabulae: The Irish Sense of Place', *Irish University Review* 18(2), 191–206.

—— (1988b), 'The Idiocy of Irish Rural Life Reviewed', *Irish Review* 5, 27–33.

Sheldrake, R. (1981), *A New Science of Life: The Hypothesis of Formative Causation* (Los Angeles, CA: Tarcher).

Singh, A. (2002) 'Speak to us, Mr Blunkett', *The Observer on Sunday*, 22 September.

Smith, A. (1988), 'The Myth of the "Modern Nation" and the Myths of Nations', *Ethnic and Racial Studies* 11(1), 1–26.

Sontag, S. (2003), 'The Idea of Europe (One More Elegy)', in *Where the Stress Falls* (London: Vintage).

Spirn, A. (1998), *The Language of Landscape* (New Haven, CT: Yale University Press).

Steinert, J. and I. Weber-Newth (2000), *Labour and Love – Deutsche in Großbritannien nach dem Zweiten Weltkrieg* (Osnabrück: Secolo).

Steinmetz, S. (1996), 'The German Churches in London, 1669–1914', in Panayi (1996), 49–72.

Stephen, W., ed. (2004), *Think Global, Act Local: The Life and Legacy of Patrick Geddes* (Edinburgh: Luath).

—— (2007), *A Vigorous Institution: The Living Legacy of Patrick Geddes* (Edinburgh: Luath).

Storry, M. and P. Childs, eds (1997), *British Cultural Identities* (London: Routledge).

Streng, P. and G. Bakay (1999), 'Volkskunde als Erlebnisagentur, oder: Von der Technik des volkskundlichen Überlebens', in Köstlin and Nikitsch (1999), 127–36.

Stummann, F. (1990), 'Europe: Projecting Different Views', *Anthropological Journal on European Cultures* 1(1), 7–11.

Svašek, M. (2007), *Anthropology, Art and Cultural Production: Histories, Themes, Perspectives* (London: Pluto).

Taylor, P. (2001), 'Which Britain? Which England? Which North?' in Morley and Robins (2001), 127–44.

Thomas, N. (1997), 'Anthropological Epistemologies', *International Social Science Journal* 153, 333–43.

Thompson, R. (1997), 'Ethnic Minorities and the Case for Collective Rights', *American Anthropologist* 99, 786–98.

Tilley, C. (2006), 'Introduction: Identity, Place, Landscape, Heritage', *Journal of Material Culture* 11(1/2), 7–32.

Tisdall, C. (2008), *Joseph Beuys: Coyote* (London: Thames & Hudson).

Tomlinson, J. (1999), *Globalization and Culture* (Cambridge: Polity).

Travis, A. (2002), 'Permits, Quotas and Tests of Allegiance', *The Guardian*, 8 February.

Trepte, H. (2004), 'Das Problem der "Hiesigen" (tutejsi) im polnisch-weißrussischen Grenzraum,' in *Krynki: Annus Albaruthenus*, 67–87.

Tschernokoshewa, E. (1998), 'Grenze als Heimat: über Reines und Gemischtes', in C. Giordano, R. Colombo Dougoud and E. Kappus, eds, *Interkulturelle Kommunikation im Nationalstaat* (Münster: Waxmann), 165–82.

Türcke, C. (2006), *Heimat: Eine Rehabilitierung* (Springe: Klampen).

Urdze, A., ed. (1991), *Das Ende des Sowjetkolonialismus: Der baltische Weg* (Reinbek: Rowohlt).

Usman, A. (2002), 'Asians Ruffled by Labour Ploy to Outwit Far Right', *Khaleeji Times*, 27 September 2002/20 Rajab 1423.

Vaišnoras, V. (1991), 'Massendeportation und bewaffneter Widerstand. Die litauische Tragödie', in Urdze (1991), 47–54.

Vallely, F. (2004), 'Singing the Boundaries: Music and Identity Politics in Northern Ireland', in Kockel and Nic Craith (2004), 129–48.

van Krieken, R. (2000), 'Citizenship and Democracy in Germany: Implications for Understanding Globalisation', in A. Vandenberg, ed., *Citizenship and Democracy in a Global Era* (Basingstoke: Palgrave Macmillan), 123–37.

Venclova, T. (2001), 'Weiße Flecken der Erinnerung', in M. Wälde, ed., *Günter Grass – Czesław Miłosz – Wisława Szymborska – Tomas Venclova: Die Zukunft der Erinnerung* (Göttingen: Steidl) 73–83.

—— (2006), *Vilnius: Eine Stadt in Europa* (Frankfurt/Main: Suhrkamp).

Verdery, K. (1997), 'The "New" Eastern Europe in an Anthropology of Europe', *American Anthropologist* 99, 715–7.

Wackwitz, S. (2005), *Ein unsichtbares Land: Familienroman* (Frankfurt/Main: Fischer).

Walters, K. (2001), *Growing God: A Guide for Spiritual Gardeners* (Mahwah: Paulist Press).

Walters, V. (2009), *The Language of Healing: Joseph Beuys and the Celtic Wor(l)d* (Ph.D. thesis, University of Ulster 2009).

Wazir, B. (2002), 'My Father Came from Pakistan and Coped with Powell and Thatcher. But Now, for the First Time, He Feels He's Not Wanted Here', *The Observer on Sunday*, 9 June.

Weber, E. (1976), *Peasants into Frenchmen: Modernization of Rural France, 1870–1914* (Stanford, CA: Stanford University Press).

Weber, M. (1972), *Wirtschaft und Gesellschaft* (Tübingen: Mohr).

Weisweiler, J. (1943), *Heimat und Herrschaft: Wesen und Ursprung eines irischen Mythus* (Halle/Saale: Niemeyer).

Werbner, P. (2002), 'Reproducing the Multicultural Nation', *Anthropology Today* 18(2), 3–4.

White, K. (2004a), *Across the Territories: Travels from Orkney to Rangiroa* (Edinburgh: Polygon).

—— (2004b), *The Wanderer and his Charts: Exploring the Fields of Vagrant Thought and Vagabond Beauty* (Edinburgh: Polygon).

—— (2006), *On the Atlantic Edge: A Geopoetics Project* (Dingwall: Sandstone).

Willke, H. (2001), *Atopia: Studien zur atopischen Gesellschaft* (Frankfurt/Main: Suhrkamp).

Wolff, S., ed. (2000), *German Minorities in Europe: Ethnic Identity and Cultural Belonging* (New York, Oxford: Berghahn).

Wright, F. (1987), *Northern Ireland: A Comparative Analysis* (Dublin: Gill & Macmillan).

Wulff, H. (2007), *Dancing at the Crossroads: Memory and Mobility in Ireland* (Oxford: Berghahn).

Zimmermann-Steinhart, P. (2003), *Europas erfolgreiche Regionen: Handlungsspielräume im innovativen Wettbewerb* (Baden-Baden: Nomos).

Zinnecker, A. (1996), *Romantik, Rock und Kamisol: Volkskunde auf dem Weg ins Dritte Reich – die Riehl-Rezeption* (Münster: Waxmann).

Zwerin, M. (1976), *A Case for the Balkanization of Practically Everyone: The New Nationalism* (London: Wildwood House).

Index

Printed in the United States
By Bookmasters